35004
35004

## H.C.S. BULLOCK

# HIS LIFE AND LOCOMOTIVES

**Kenneth Allan Bullock**

**Revised, Updated & Edited by Bob Bullock**

*No. 2005 'Silver Jubilee' at Bramber Castle, Steyning in the 1930s*

*This revised edition of 'H.C.S. Bullock: His Life & Locomotives' is reprinted from the 1987 edition with the permission of The Heywood Society and Andrew Neale, Plateway Press*

## Author's Notes

*Having inherited my father's passion for steam engines but not, alas, his ingenuity, I feel it is my humble duty to record in this book some of his remarkable achievements, not least of which was the design, construction and reconstruction of no fewer than seventeen 10¼-inch gauge steam locomotives during the depression years of the 1930s. Most of his locomotives still survive today, and a fair percentage are still operating.*

*My thanks must go to John Hall-Craggs and Brian Hollingsworth for their great help and encouragement, and to Freda Ridgett for her help with the typing; also to my son Keith for his diagrams and moral support. I must also acknowledge the debt I owe to John Hall-Craggs, Brian Hollingsworth, Matthew Kerr, James Nutty and Simon Townsend for generously lending from their private collections many precious photographs for the book.*

*Kenneth Allan Bullock, 1987*

*Ken Bullock at the controls of 'The Monarch' on the last day of operation at the Meadow End Railway, Brockenhurst, in 1966. His son Keith stands to the left*

**H.C.S. Bullock: His Life and Locomotives**
Republished in 2017 by *A to B Books*, 40 Manor Road, Dorchester DT1 2AX
01305 259998
atob@atob.org.uk
www.atob.org.uk

Printed in the United Kingdom by Henry Ling Ltd, at the Dorset Press, Dorchester DT1 1HD
*All photographs are from Ken Bullock's private collection unless otherwise stated. All maps by A to B Books*


First edition 1987, ISBN 0-9511108-9-6
Revised second edition April 2017
**ISBN 978-0-9575651-2-8**

**Acknowledgements** *I am indebted to the following for their help, the loan of photographs and materials and encouragement: Keith Bullock, Simon Townsend, James Nutty, Paul Stileman, Clive Upton, Peter Scott, Rob Bezley, Rob Hart, John Kerr, Ken Bean and Andrew Neale. Special thanks to David Henshaw of A to B Books for coping with my many updates and corrections without complaint*

*This book is dedicated to H.C.S Bullock, locomotive engineer* PHOTO:: Holder Family Collection

## Introduction to the second edition

*Ken Bullock's original 1987 book about his father gave a wonderful insight into the man, and a detailed look at the locomotives he built. It was followed in 1993 by 'The Surrey Border & Camberley Railway' by Peter Mitchell, Simon Townsend and Malcolm Shelmerdine, which carried on from Ken's book, with more details of the locomotives and a superb account of the Foxhill and Surrey Border & Camberley Railways.*

*When I decided that it was time Ken's book was brought up to date, as so much had happened in the intervening 29 years, I concluded that there was no point in simply repeating what Ken had already written. I was fortunate in becoming very good friends with Ken, and spent a lot of time with him at Eastleigh Lakeside Railway, driving his father's engines, and at his home, talking about the railways and looking at his photograph collection. Ken's son Keith very kindly allowed me full access to his father's photographs, and so the update to the book became in part a photographic section.*

*The coming of the home computer and image processing programmes such as Adobe Photoshop, and more recently Lightroom, meant that many of Ken's photographs which were not suitable for publication in 1987 could now be scanned and restored to something approaching their original condition in many cases. However, I have included quite a few images which are not of a very high standard because of damage - or simply through being too poor in the first place - because I felt they were of historical significance and would be of interest to enthusiasts and miniature railway historians.*

*During the early days of locomotive manufacture, and during the construction and operation of the Foxhill and Farnborough Miniature Railways, many photographs were taken, but as things became more hectic - especially at The Olives - there were, unfortunately, very few photographs.*

*The new, and newly restored, photographs are all from Ken Bullock's private collection, unless otherwise credited.*

*And finally, to avoid the inevitable confusion and speculation, I am not a relative of Ken's, or indeed related to his line of the Bullock clan in any way, as far as I know!*

*Bob Bullock, 2017*

*No. 2 'Wendy' and No. 2006 'Edward VIII' waiting in the return loop at Hawley on the Foxhill Railway*

*H.C.S Bullock gives No. 2006 'Edward VIII' a polish with 'Blue Billy' before starting the working day at Fox Hill. These must have been happy times, with his own railway, and the prospect of expansion with the arrival of a partner, Alexander Kinloch, who was prepared to invest a lot of money in the venture*

# Contents

# I
# Peace & War
# 1889-1922

My father's interest in steam locomotives became apparent when he was old enough to bite the shape of an engine out of his bread and butter! He was born in 1889 and spent much of his childhood with his two brothers in the Wiltshire village of Box, where his father was a signalman on the Great Western Railway at Box station, working in the famous 'Box Signal Box'. Like most young lads, the brothers were not beyond getting up to a certain amount of mischief and I recall my father telling us of some of the pranks, which didn't always work out the way they were planned.

One such occasion happened after the Council had left a wooden tar barrel at the top of the hill in the village in readiness for some road repairs. Opposite lived a rather grumpy local dignitary. The boys, therefore, decided to leave this gentleman the tar barrel outside the front door as a present. Unfortunately, en route to its destination, the barrel ran out of control and plunged down the hill towards the Northy Arms Hotel, on the Bath Road, finally smashing itself against a stone wall halfway down the hill. Evidence of the incident could still be seen as late as 1975!

After spending a short time as a cleaner at the GWR's Swindon works, Father moved into lodgings at Pilning, Gloucestershire, and obtained work at the pumping station at Speedwell Colliery. He soon showed an ability in drawing, and produced some fine sketches of the saddle-tank locomotive, the winding gear and the boiler house at the colliery.

In February 1909, Father enrolled in the 4th Gloucester Regiment, transferring to the Royal Flying Corps in June 1912. He was then posted to Farnborough in Hampshire, with No.2 Squadron, moving to Montrose, Scotland, in February 1913.

During these early pioneering days of the R.F.C. many activities of an inventive and experimental nature were taking place, not least of which was a continuous effort to increase the flying range of the aeroplane. It was during this period that a 'Length of Flight Time' record was achieved. Quoting from 'The War in the Air, volume 1' by Walter Rayleigh, 'In the front seat of a B.E. machine, 1st Class Air Mechanic H.C.S Bullock fitted a petrol tank of his own design, estimated to give at least eight hours fuel for the 70hp Renault engine. On 22nd November 1913, Captain Longcroft took off in this aircraft and flew from Montrose to Portsmouth, circling inland, finally landing at Farnborough in seven hours 20 minutes non-stop'. My father built a fine model of the aircraft concerned.

At the outbreak of World War I No.2 Squadron were sent to France and soon saw action. Whilst my father was flying low over enemy lines as Observer to this same Captain Longcroft their aircraft came under some heavy rifle fire and was subsequently forced to land through loss of oil pressure. It was soon discovered that a bullet had damaged a feed pipe causing serious loss of oil. By using strips torn from his shirt sleeve, Father was able to effect a temporary repair and restart the engine, taking off virtually in site of a German patrol which had been sent to capture them.

My father got married in Montrose during 1915. His bride was a girl from the town called Helen Alexander. They had more time together than most newly-wed couples in those terrible days, because my father received shrapnel wounds on Christmas Day and he was sent home for six month's sick leave. After this he returned to France for a further tour of duty with No.8 Squadron.

At this time, my father was involved with the problems of night flying, which was causing so many casualties without compensating success that the authorities were considering prohibiting it altogether.

My father played an important part in the development of a device or apparatus - it isn't clear exactly what this was - which would assist pilots in this hazardous operation. His work played a significant part in Lieutenant W. Robinson's success in bringing down the Zeppelin L22 at 2.45am on 3rd September 1916.

My father was eventually posted home in 1917, and he served at the Central Flying School, the Aerial Gunnery Schools at Hythe and Loch Doon and the Experimental Station at Orfordness. While at Hythe, Father's efforts in an experimental capacity to bring about a successful launching of an aerial torpedo were also noted. During his time at Orfordness, my father was sent to Felixstowe to reorganise the workshops on the coast which catered for the repair and maintenance of small naval craft.

Of the 21 vessels which comprised the normal strength of the establishment, only a handful were seaworthy when he arrived. By setting an example (in borrowed overalls!)

and personally involving himself in the work, he was able to bring the unit up to full strength in a comparatively short time. There is a report of my father pointing out to higher authority that an expensive and elaborate scheme to drain some land to the north of the base at Montrose was a non-starter for various reasons. Being a junior officer his technical advice fell on deaf ears among the military hierarchy, but in the end - after a great deal of expense and wasted time - the scheme had to be abandoned for the reasons my father had put in his report.

After a distinguished career, my father left the service in 1919 with the rank of captain. He received the MBE in January 1918, followed by a mention in despatches in March. He had been awarded the Military Medal in January 1916 for his exploit with Captain Longcroft, and in that connection it must be said that he was never without a sense of humour.

When he was summoned to Buckingham Palace to receive his decoration, the overnight train from Scotland was so late that it left him no time to shave. At the investiture ceremony, King George V asked him where he had served last, and my father thought he had said 'Where did you shave last?', to which he replied, 'Scotland Sir.' The King looked bewildered, probably as the Military Medal was awarded for gallantry in the field!

After leaving the service, my father moved south to join his two brothers at Palmerston Road, Boscombe, Bournemouth, where they invested an impressive £15,000 (£620,000 today) in a small factory to design and build a light car called the Palmerston, which appeared in 1920. Unusually - at a time when light cars were often little more

*A rare photo of the Bullock family riding - or at least sitting - in a Palmerston at Bournemouth. Mrs Bullock is driving, with a dapper-looking H.C.S and an unidentified junior Bullock*

*Judging by these advertisements from 1920 (above) and 1921 (below) the Palmerston had some impressively modern features, but these were hard times, and there were plenty of other young ex-servicemen producing similar machines* PHOTO:: The Automobile Oct 2008

than kits built from proprietary parts - the Bullock brothers had their own foundry, and seem to have made many of the parts themselves.

The Palmerston was shown at Olympia, and spoken of as 'a machine of simple and pleasing appearance', which was claimed to be the first light car of its kind. It was reviewed in both *Light Car & Cyclecar* (26th June 1920), and *The Autocar* (13th November 1920).

The cost of this baby car was £250, and it was a two-seater with a 'dickie' seat. It was driven by a 680cc Coventry Victor water-cooled engine, and could return 55mpg. In 1921, a slightly enlarged and rather more powerful model was introduced, with a 1018cc engine. This expansion seems to have brought the venture to an end, because the company failed in February 1921, leaving debts of £4,500 (£183,000 today). The business was sold, and both models were still available in 1922, when they were included in *The Autocar's* 'Illustrated Review of 1922 Models'. Despite the publicity, only a few cars were produced - none of which seem to have survived - and it must be said that fathers' sojourn in the car manufacturing business was not a complete success. But a move to Farnborough was in the offing, and that would bring success in a very different field.

# 2
# Locomotive Construction Begins 1923-1933

*5-inch gauge 4-6-0 No. 6000 'King George V', built at Fowler Road in 1927*

In 1923, my father joined the Air Ministry, and was sent from London to work in the Design Department at the Royal Aircraft Establishment at Farnborough. He was given staff quarters in Fowler Road, Cove, and this proved to be his opportunity to start locomotive building. The first locomotive was a 2-4-2 wooden toy with handles attached to the cab sides to assist the author - then aged 15 months - in the art of walking!

Two proper 4-6-0 locomotives followed, one each for my brother and me, built to the three-inch gauge. They were fired by methylated spirit held in a tray under the boiler, which was made from a cocoa tin. Serious construction began in 1924, and over the next few years several 3½- and 5-inch gauge engines were built:

| | | |
|---|---|---|
| **4-2-2** | ***Duke of Connaught*** | *GWR Achilles class* |
| **0-4-0T** | | *Freelance design* |
| **4-2-2** | ***Black Prince*** | *Rebuilt Achilles class* |
| **4-4-0** | ***Bulldog*** | *GWR Bulldog class* |
| **4-6-0** | ***Pendennis Castle*** | *GWR Castle class* |
| **4-6-0** | ***King George V*** | *GWR King class* |
| **4-6-2** | ***Lady Hazel June*** | *GWR-style loco* |

*'Lady Hazel June'*, shown on page 12 before completion, was intended as a modernised version of the GWR 4-6-2 *'The Great Bear'*. She was the prototype for my father's subsequent larger locomotives, including *'Duke of Connaught'*, *'Pendennis Castle'* and *'King George V'*, which all gained major awards at Model Engineers exhibitions, the Castle winning the bronze medal in 1927, and *'King George'* the silver in 1928. All the locomotives, together

*LEFT: 5-inch gauge GWR-style 0-4-0T. Ken Bullock is too interested in driving to look at the camera!*
*RIGHT: GWR 'Achilles' class 4-2-2 'Black Prince' hauls a load of young Bullocks away from the platform*

with a number of steam-driven feed pumps which my father had made and patented, were sold to Captain Holder of 'Keeping' at Bucklers Hard in the New Forest.

In 1932, No.1001, *'The Monarch'* was completed, this being the first Bullock 10¼-inch gauge locomotive. Once it had been delivered to Captain Holder, the ensuing weeks were devoted to dismantling the workshop at Fowler Road, and moving everything to our new house, 'The Olives', in Prospect Avenue, Farnborough. The Olives was a spacious four-bedroomed dwelling, jointly designed by Mr Hilder - the manager of a local building firm - and my father.

The drawing office, where my father was to spend most of his time, was light and roomy, and overlooked the copper beeches of Prospect Wood. The garden at the rear was 120 feet in length, and on the right-hand side, a large workshop was built with two

*3½-inch gauge locomotive 'Pendennis Castle'. Awarded the Bronze Medal at the Model Engineer Exhibition in 1927*

*'Lady Hazel June' under construction at Fowler Road. Although only 5-inch gauge, this Great Western-style locomotive showed clearly the direction H.C.S Bullock was heading in the late 1920s*

*The first 10¼-inch gauge locomotive, No.1001 'The Monarch' with frames complete and one cylinder fitted, outside the workshop at Fowler Road in 1930. In these early days, when locomotive manufacture was still an enjoyable hobby, Bullock took plenty of photographs. Construction of later machines was less well recorded*

*Some months later, still waiting for her boiler, but 'The Monarch' is coming along nicely*

*No. 1001 'The Monarch' as delivered to Captain Holder, still with Baker valve gear* PHOTOS:: Holder Family Collection

*There seem to have been a few teething problems, and 'The Monarch' (soon renamed 'Audrey') was partially rebuilt by Bullock at Keeping in 1936, emerging with a new boiler, Walschaerts valve gear, a sand dome, and steam-driven boiler feed-pump*

*Close-up of the sanding pipework, Walschaerts valve gear and steam-driven pump. Changes often have a slightly untidy look, but these were neatly integrated*

*J.N. Maskelyne, Editor of 'Model Engineer', drives 'Audrey' across the lawn at 'Keeping' in 1935*
PHOTO: Holder Family Collection

ABOVE: There don't appear to be any photographs taken inside the workshop at The Olives, but a few tantalising glimpses from the outside. This is 'Edward VIII', apparently close to completion. Looking at what is effectively a large garden shed, it's hard to believe such magnificent machines were created here

RIGHT: In the light of experience, the locomotive works at The Olives was carefully laid out to make locomotive movements easier, and even allow for limited testing prior to despatch. The precise track layout is still open to some doubt, but the first track definitely ran from the locomotive loading point to the workshop (as above). As the business expanded, the car was evicted from the garage, which was then served by a second siding, with another running down to the turntable, which would later be positioned in the left background of the photo above. This plan shows the most extensive layout, before the turntable was moved to Fox Hill

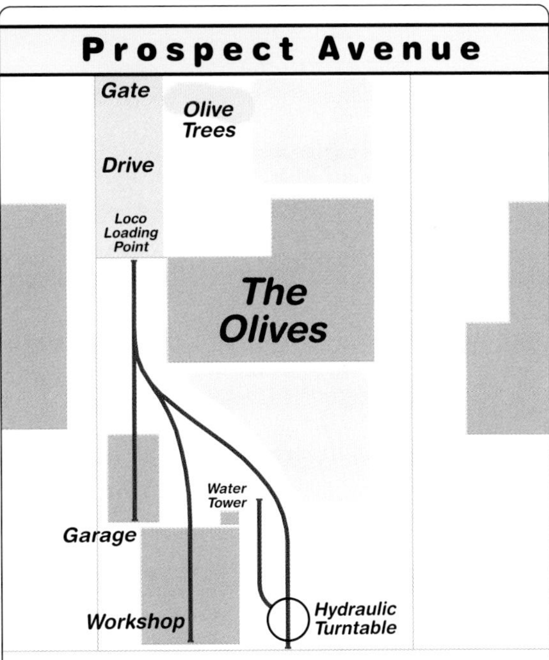

ground-level tracks concreted into the floor. These led out to a set of points in the drive, becoming a single track which curved towards the house, then ran parallel to the wall, terminating in line with the front of the building. A pair of heavy gates concealed all this from Prospect Avenue.

Looking back, I feel the days at Fowler Road were a happy and settled time for my father, as he seemed to have more time to spend with us children. Indeed, as the family increased, my father found a need for a bigger car than the drophead Talbot with its Dickie seat that had served as the family car.

My father solved the accommodation problem by removing the original body and fitting a saloon type, which was relatively easy to do in those days when cars had separate chassis. He made the body himself by laminating plywood, and sealing it with an

*'The Empress' was built in 1933 for Mr H. G. Cookson's private railway at Billingshurst, Sussex. 'The Monarch' had in many ways been a test-bed for Bullock's large-scale locomotives, and many of the lessons learnt resulted in modifications incorporated into 'The Empress', Seen here outside the workshop at 'The Olives', 'The Empress' was the machine that the following 1000-series locomotives were based on*

ABOVE: It's not clear whether - and to what extent - builds overlapped, but No. 1003 'Western Queen' must have been underway before 'The Empress' was dispatched

RIGHT: There were certainly lots of smaller jobs while the big Pacifics were under construction. This is the famous 9 1/2-inch gauge London Brighton & South Coast Railway's Stroudly 0-4-2T 'Ranmore', in for overhaul. Note the Bullock family Citroen car in the garage and - at this early stage - the simple track layout

*No. 1002 'The Empress' before leaving The Olives, telling the world she really should have been a Great Western engine!*

*Bullock in the driving seat of 'The Empress' on Mr Cookson's line at Billingshurst. A few modifications have been made: a taller chimney, a different dome cover, and 'Great Western' has been removed from the tender. Mr Cookson was a Southern man!*

outer fabric skin. He secured the front of the roof to the existing windscreen by utilising the two vertical threaded stays and screwing down wing nuts in concealed apertures.

However, on the road, the wood soon began to squeak against the brass windscreen frame, so packing was placed between the two faces. This packing usually consisted of halfpennies and sometimes even pennies. It wasn't long before the younger members of the family discovered this hidden treasure trove, resulting in more frequent trips to the sweet shop. Dad continued to replace the packing whenever it 'dropped out'. I like to think that he knew perfectly well what was happening to it!

The journeys we made through the Forest to visit Captain Holder always seemed an adventure in the old Talbot, as there was at least a 12-mile gap between filling stations, but how reassuring it was to pass an AA patrolman on his motorcycle and sidecar, receiving the customary salute when he saw the AA badge on the radiator.

4-6-2 No.1002 *'The Empress'* was the next locomotive to be built at Farnborough. She took over 12 months to complete, being finished in 1933 and sold to a Mr H. G. Cookson of Billingshurst, Sussex. As with No.1001, she was finished in GWR green with a copper-top chimney. *'The Empress'* was taken away on a trailer towed behind a big black and green Morris 18hp saloon.

*The 'Empress' was later returned to The Olives and re-gauged to 9 1/2-inch gauge for Vere Burgoyne's railways at Crowthorne and Spring Lanes. For some reason (she had been properly tested at the new Foxhill Railway), she was steamed at The Olives, and this photo was taken with most of the family aboard, and Mrs Bullock driving. It wouldn't have been much of a ride, of course, because the locomotive is almost on the turntable (it may already have been removed by this time), and there is barely a train length's of track out of sight behind the workshop*

# 3
# Full-time Engineer 1934-1935

*Built in 1934, 'Western Queen' was virtually identical to 'The Empress' with piston-valve cylinders and Baker valve gear. Finished in GWR green and lettered 'Great Western', she is shown when brand new outside the workshops at The Olives*

After the completion of No. 1002, my father resigned from the Royal Aircraft Establishment to concentrate on the design, construction and reconstruction of 10¼-inch gauge locomotives. The single-wheeler from the Holder's earlier line at Pitmaston Moor Green in Birmingham, was at Keeping at about this time, and was brought to Farnborough to be rebuilt with a larger steel boiler in place of the original copper one.

The locomotive was modelled on Patrick Stirling's famous GNR single-wheeler, and was built by Grimshaw, some time before 1910. The rebuild took place alongside the third Pacific then under construction, No. 1003 *'Western Queen'*.

*'Western Queen' was hired to a Mr Ballance who opened a new railway at Burnham-on-Sea, Somerset in 1934. She ran there for two seasons, looking very smart with polished boiler bands, but on visiting Burnham, Bullock discovered that she was not being mechanically well maintained, and sand had entered the lubricator. The locomotive, together with the Stirling single-wheeler, were taken to Farnborough for repairs, and the Burnham Miniature Railway subsequently closed*

*It's not clear how useful the Stirling single-wheeler proved to be at Burnham-on-Sea. Like all machines of its kind, it was somewhat wanting in the traction department, but a very elegant locomotive* PHOTO:: Bob Bullock Collection

When both these jobs were completed in 1934, No.1003 and No. 1, the 4-2-2, were hired to Mr Ballance of Burnham-on-Sea, Somerset. No.1003 was of similar dimensions to the other 4-6-2s, but with two injectors, and detail alterations to boiler fittings.

A second locomotive for Captain Holder, a 4-6-4T numbered 1004 to be named *'Mary'* had reached a state of near completion towards the end of 1934. Alongside No.1004 in the workshop was the old Bassett-Lowke 4-4-2 *'Peter Pan'* of Wembley Exhibition fame, which was undergoing a rebuild, including regauging from $9\frac{1}{2}$- to $10\frac{1}{4}$-inch gauge. It had come up from Keeping during the summer.

*H.C.S didn't have many clients in 1934, but they were keeping him busy. This is Bassett-Lowke 4-4-2 'Peter Pan' on Captain Holder's Railway, rebuilt and re-gauged to $10\frac{1}{4}$-inch* PHOTO:: Holder Family Collection

My father's policy of using standard parts for his locomotives, wherever possible, i.e. wheel and cylinder castings, axle-boxes, horn blocks and so on, enabled him to cut down on the labour and expense of having to produce several different patterns and drawings for the chassis. In fact, the engine parts of his locomotives were virtually identical up to No. 1004.

*After returning to Farnborough, 'Western Queen' was repaired, repainted black, lettered 'LNER' and used during the construction of Bullock's railway at Foxhill. Here - with Ken Bullock and his father in attendance - she is being loaded onto the trailer at The Olives, presumably en route to the work site just half a mile up the road. Even with a modern trailer, moving such a large locomotive would be difficult, and it must have been a hair-raising operation for a man-and-boy team using a single-axle trailer. This picture is also interesting because it shows where loading took place, on the west side of the house*

*4-6-4T No. 1004 'Mary' was built in 1934 for Captain Holder's railway at Keeping. She was the last of the 1000-series locomotives, and beneath the side tanks, mechanically identical* PHOTO:: Holder Family Collection

Nevertheless, things were getting somewhat hectic at Prospect Avenue during this period. Mr Cookson had placed an order for an additional locomotive, while the now rebuilt *'Peter Pan'* had been replaced by the other 4-4-2 *'John Terence'*, also from Captain Holder. There was also an order for a six-coupled $9^{1}/_{2}$-inch gauge tank engine from the Spring Lanes Light Railway at Bracknell, owned by Vere Burgoyne. All these jobs were dealt with at the same time.

First to emerge in the spring of 1935 was 4-6-2 No. 2005 *'Princess Marina'*, the locomotive for Mr Cookson. It had slide-valve instead of piston-valve cylinders, of 5-inch stroke by $3^{1}/_{2}$-inch bore and 14-inch diameter driving wheels. In addition to high- and low-pressure injectors, No. 2005 had a feed pump fitted on the right-hand running plate. This engine was heavier, and with a working pressure of 140psi, was more powerful than any of the previous ones. It would turn out to be the first of Bullock's second standard design.

As was normal practise with all the locomotives turned out, No. 2005 was blocked up, with it's driving wheels well clear of the ground and run in at full working pressure for a day, alternating between full forward and full back gear. A partial application of the steam brake from time to time to stimulate load would greatly increase the sound of the exhaust beat, so I was glad when my stint to look after the engine came round, although Dad was never far away.

The name *'Princess Marina'* was changed to *'Silver Jubilee'* prior to leaving Prospect Avenue, to mark the celebrations held in May 1935 to mark King George V's 25th year on the throne. Such was the patriotic fervour of the time that soon after arrival at

*'Mary' at Keeping in 1934. Water and coal capacity were always limited, and the locomotive was later despoiled with a tender. In this photo, it seems to be carrying an additional coal supply in a box in the first wagon* PHOTO:: Holder Family Collection

Billingshurst, Mr Cookson had the oil cups, clack boxes, hand-rails and other detail fittings removed and chromium-plated. *'Silver Jubilee'* looked resplendent double-headed with *'The Empress'* on Mr Cookson's picturesque private line in West Sussex. Even before No. 2005 had been completed, however, work was well underway on the next 4-6-2, No. 2006 *'Edward VIII'*.

*'John Terence'* had been satisfactorily steam-tested after its rebuild, and duly returned to Captain Holder, and the 9½-inch gauge tank engine from Mr Burgoyne's line had also been finished. Apart from a diminutive orange-painted 0-4-2T in the workshop for valve setting, *'Edward VIII'* was temporarily the sole occupant of the workshop.

The workload at Prospect Avenue had been eased slightly by sending the wheel and cylinder castings out for machining to a light engineering works at North Camp near Aldershot. This was owned by a Mr Panting, known locally as 'Digger'. My brother Charles and I were also involved in keeping the workshop floor uncluttered, as well as

*Three 'foreign' engines overhauled or rebuilt by H.C.S Bullock*

*RIGHT: A very poor photograph, but full of interest. 'John Terence' in action on Captain Holder's railway. Mrs Bullock is on the train*

*LEFT: A 7¼-inch gauge miniature of Great Central Railway 4-6-0 No. 6097 'Immingham', overhauled at Fowler Road in 1931. According to Ken, the locomotive was subsequently advertised for sale for £100. The fact that he remembered this detail half a century later suggests that it was a topic of family conversation at the time. It's tempting to conclude that H.C.S. might not have been charging enough for his labours*

*RIGHT: The six-coupled tank engine rebuilt by Bullock for Mr Burgoyne's 9½-inch gauge Spring Lanes Light Railway at Bracknell, Berkshire, in 1935*

*ABOVE: 4-6-2 No. 2005 'Silver Jubilee' posed for the now familiar 'works' photo at The Olives. This was the first of Bullock's 2000-series locomotives, with slide valve cylinders and 14-inch driving wheels. Built for Mr Cookson's private railway at Billingshurst in Sussex, it would later run on the Surrey Border & Camberley. 'Silver Jubilee' was an impressively finished locomotive, with such details as lamp brackets, steps and polished smokebox door hinges, although the look is slightly marred by the roughly lagged cylinder steam pipes*

*LEFT: The curved shed road at The Olives must have been wide to gauge. Wisely, Bullock used broad treads and deep flanges!*

*BELOW: The auxiliary hand-operated boiler feed pump*

such tasks as drilling pilot holes, de-burring axle-boxes after milling, and turning and facing buffers on the centre lathe.

Another new arrival in the workshop was North Eastern Class R 4-4-0 No. 2015, sent in for rebuilding in the style of an LNER 'Super-Claud' D16, to be named *'Wendy'*. The locomotive number in this case was the number inherited from the prototype rather than my father's own number. It would not have fitted his ingenious scheme whereby the left-hand digit represented the number of the design, while the right-hand ones represented the 'works number' of the locomotive in sequence. For example, No. 1003 *'Western Queen'* was the third Bullock 10¼-inch locomotive of the first design.

Towards the end of the running season, my father made a trip to see the line at Burnham-on-Sea, and discovered that sand had been permitted to enter *'Western Queen's'* lubricator. He made immediate arrangements for No. 1003 and the little single-wheeler to be returned to Farnborough. This incident prompted my father to renew his efforts to acquire a suitable piece of land where he could lay a quarter-mile or so of track for his own purposes. This was achieved late in 1935.

*4-6-2 No. 2005 'Silver Jubilee' in action on Mr Cookson's railway at Billingshurst in Sussex*

*4-4-0 No. 2015, based on the North Eastern Railway Class R, later LNER Class 20) passed through The Olives for rebuilding. It was an increasingly busy time, with overhauls of other locomotives slotted in between the many new locomotive orders then on the books. Confusingly, No. 2015 would be given the name 'Wendy', but she was not the similar 'Wendy', No. 2, that worked at Foxhill and briefly on the Surrey Border & Camberley*

A nearby dairy farmer called 'Master' Bill Ratcliffe owned some pasture land about half a mile from Prospect Avenue, which was known as 'Foxhills'. After some discussions, an arrangement was made between my father and Mr Ratcliffe to lease a strip of land on the lower part of the meadows to a point terminating in some woods, which we knew as Lye Copse.

During the autumn of 1935, staff were employed to erect a shed about 30ft long by 20ft wide, at the closest point to the farm track. A water supply complete with drainage facilities was provided.

It was at this point that an immaculate *'Western Queen'*, now in unlined black livery and with a squat cast-iron chimney, was delivered on a purpose-built two-wheeled trailer towed gingerly behind our big Citroen saloon car. Soon 12lb/yd rail began arriving from Messrs Francis Theakston and construction started of platforms for two roads.

*Another view of the Bullock's trailer, this time behind the family Citroen. It wasn't very far to Fox Hill, which is perhaps fortunate. This is No. 1003 'Western Queen' arriving at the new railway. Most of the other equipment would have been carried in the 3-ton lorry*

The rail sections were

*Construction work at Fox Hill, with an engineer's train about to depart, carrying pre-fabricated track panels. Note the new ballast, tank wagon and impressive signal gantry*

built up on hardwood sleepers and after assembly loaded into flat bogie wagons. Then they were propelled by *'Western Queen'* to the work site and laid onto a prepared track bed. The bogie wagons would later be adapted to form some of the passenger rolling stock of the railway.

A turntable was built at Prospect Avenue and set up in a prepared pit. It was experimentally operated by hydraulics, as was one set of points, but my father considered the hand-operated pump too slow in practise, so he abandoned the experiment.

The points and the turntable - which was to serve three roads - were transported to the line in a very antiquated 3-ton Morris lorry which had been bought to transport equipment from the workshop. This was also used for collecting wheel castings from Wallace & Stevens, the raw castings now going direct to Mr Cookson for machining and pressing onto the axles, ready for bogie assembly at Fox Hill.

This work was very rewarding, but a bit tedious, as each axlebox contained 15 rollers, and there were eight axle boxes to pack with grease and load with bearings, plus eight covers with spring washers and set screws for each coach. Assembly certainly made our hands cold on those frosty winter mornings out on the platforms.

An all-steel bogie ballast wagon and 60-gallon four-wheeled water tank wagon were built in the workshop and placed in service immediately. An all-weather saloon coach, designed by my father, was also sent to the line. This coach was fitted with padded seats and finished in chocolate and cream livery. Two similar coaches were to follow, although these were built by Baxter's Coach Works of Bagshot, Surrey.

# 4
# Partnership
# 1936-1937

*Fox Hill Station. Left to right, 'Princess Elizabeth', 'Edward VIII', 'Harvester' and 'Western Queen'* PHOTO: D C Thompson & Co

Alexander Kinloch, a merchant banker of 20 Eaton Square, London, had been a frequent visitor to The Olives for several weeks up to this period. It ultimately transpired that a more elaborate scheme, financed largely by Mr Kinloch, was then undertaken, with my father's role being that of 'Chief Mechanical Engineer & General Manager'.

Mr Ratcliffe agreed to the line being extended through the copse and into more meadow land to the Blackwater River, and beyond to his boundary, which was about 100 yards east of the Fox Inn at Hawley. Mr Ratcliffe stipulated that the line be fully fenced and that a barrier be placed across it outside running hours, between the copse and the Fox Hill side, to prevent cattle straying. This barrier was to figure in a dramatic turn of events later in the history of the railway.

Extra staff were employed to push on with the line through Lye Copse. Lighter 9lb/yd rail was now being used and a loop was incorporated with spring-loaded points at each end. Several embankments had to be formed due to the boggy and undulating nature of the ground through the woods. It was not a very hospitable place to be

*The boiler for No. 2006 'Edward VIII' under construction at Goodhand Bros in Gillingham, Kent*

during the winter, and doubtless the men were glad to be out on the level ground again.

In due course, the Cove Brook was bridged with rolled steel joists supported at each end by transoms secured to wooden pilings driven into the bank. An emergency water tower was erected alongside the track on the Farnborough side of the river and a semi-rotary hand pump was mounted nearby to supply the water. The track continued over the bridge to a stiff pull up at 1:70 to Hawley Station buffer stops, a few yards from the Fox Inn boundary. Three lines served the two platforms.

The distance from Fox Hill to Hawley was just under 1 1/4 miles. A reversing loop of 90ft radius was added in 1937, so avoiding running round, as well as tender-first running.

*ABOVE: H.C.S. obviously decided that the Foxhill Miniature Railway made a better backdrop for works photos than the workshop at The Olives! No. 2006 'Edward VIII' makes a fine picture in the headshunt at Hawley*

*BELOW: 'Edward VIII' on the turntable at Fox Hill. The loco shed is behind and the water tower on the right*

At Prospect Avenue, No. 2006 *'Edward VIII'*, now completed, was steamed to full pressure (140psi) to check the 'pop' safety valves, the first locomotive to be so fitted. There was also a five-element Swindon-type superheater, as well as sanding gear. The whistle lay horizontally along the top of the Belpaire firebox and had a very distinctive tone. In most other respects, No. 2006 was similar to No. 2005, but more powerful, with cylinders of four-inch bore and five-inch stroke. Both locomotives were fitted with boilers made by Goodhand Bros of Gillingham, Kent.

Two 0-6-0 tank engines were under construction during the winter of 1935, Nos. 3007 and 3008, with pannier and side tanks respectively. Both had inside cylinders of 2¼-inch by 4-inch and 12-inch driving wheels. The working pressure was 125psi. No. 3008 was fitted with vacuum brakes and was the only Bullock locomotive so fitted.

One of several rebuilds undertaken by Bullock at this time was 4-4-0 No. 2 *'Wendy'*, which was the second locomotive to arrive at Fox Hill. Her presence enabled *'Western Queen'* to have the luxury of an odd day off. *'Wendy'* was followed a week or two later by *'Edward VIII'*, the running shed having been considerably improved and made more secure to accommodate three locomotives.

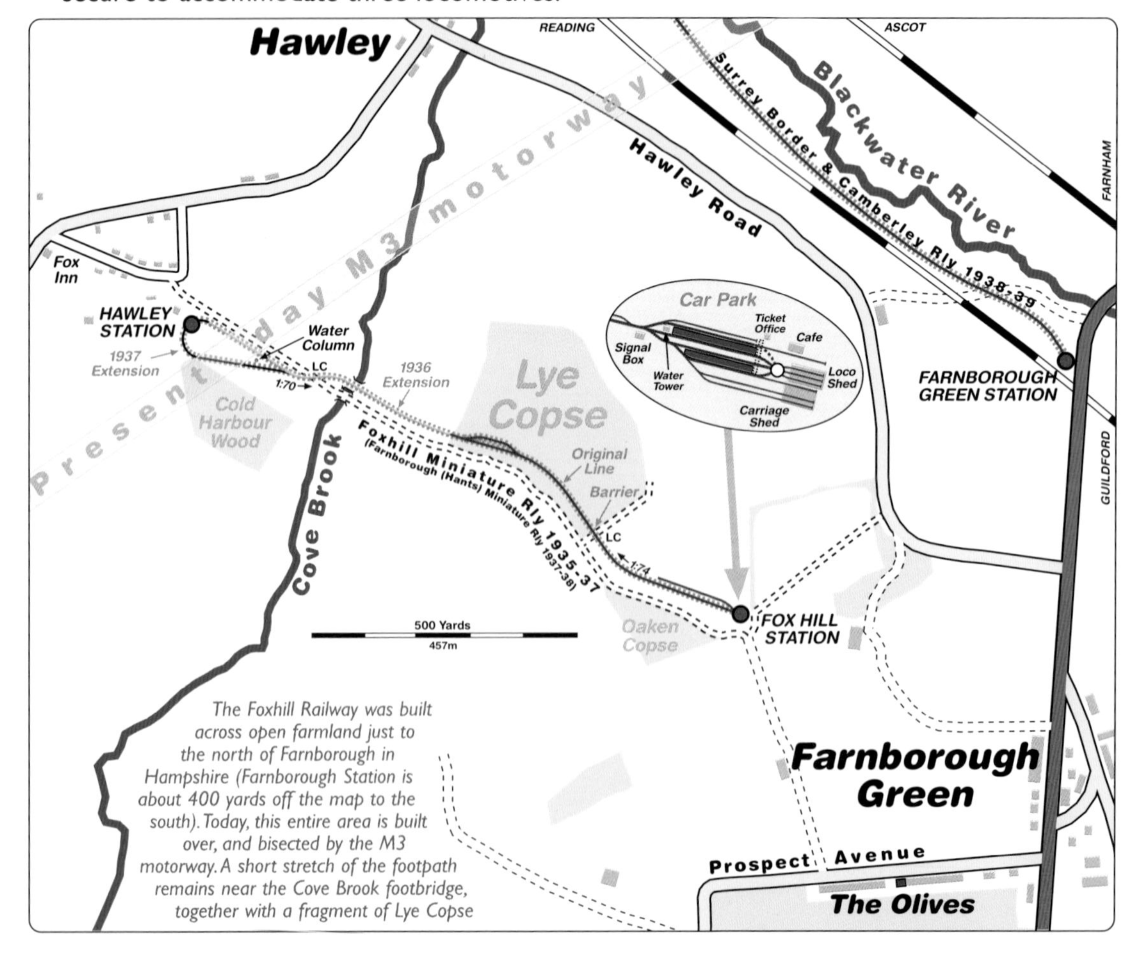


*The Foxhill Railway was built across open farmland just to the north of Farnborough in Hampshire (Farnborough Station is about 400 yards off the map to the south). Today, this entire area is built over, and bisected by the M3 motorway. A short stretch of the footpath remains near the Cove Brook footbridge, together with a fragment of Lye Copse*

Mr Kinloch and his close friend and confidant, Dr. Bernard, were frequent visitors to the railway, arriving from London on Saturday and staying in Bagshot overnight, allowing them a full day's driving on the Sunday. Mr Kinloch appeared to be uneasy whenever he drove *'Edward VIII'*, preferring the more docile *'Western Queen'*. No. 2006 had a tendency to prime, and of course the 'pop' safety valves were inclined occasionally to startle the unsuspecting driver, to say the least!

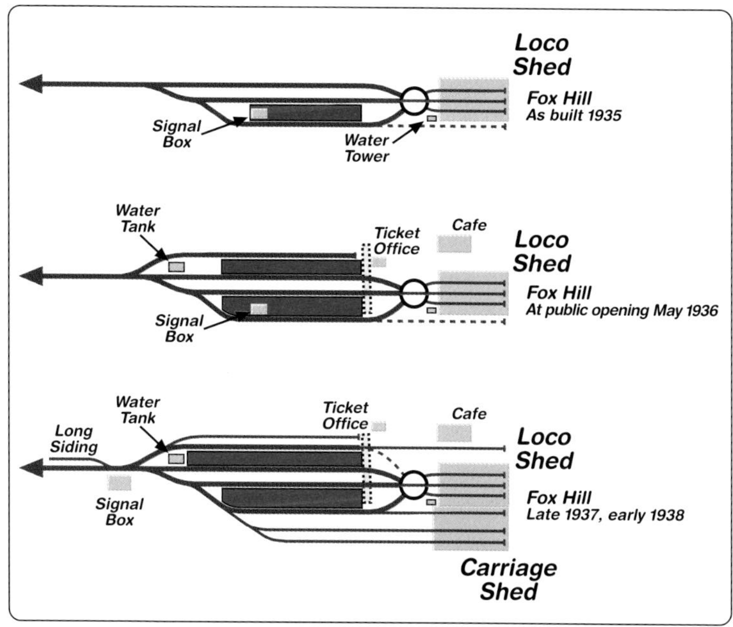


*The layout at Fox Hill was in a state of constant flux throughout its short life. The sidings shown here are thought to be correct, but others may have come and gone. The final short-lived layout is post-H.C.S Bullock*

In due course, the 0-6-0 pannier tank No. 3007 was pushed outside the workshops at Prospect Avenue for her first steaming. Unfortunately, both injectors gave trouble, and eventually the fire had to be drawn. The problem, I seem to recall, had something to do with the water in the pannier tanks becoming too warm for the injectors to pick up without repeatedly ceasing to feed. This was put right shortly afterwards, and the locomotive was loaded onto the trailer and transported to Fox Hill behind the old three-tonner.

In the new year, my father made another complete change from the two tank locos by designing a replica A3-class LNER 4-6-2. The engine took its number from its prototype, No. 2573 *'Harvester'*, instead of following the system used with the other locomotives. The leading particulars were:

**Cylinders** *Two, of 3 3/8" bore x 4 3/4", with piston-valves*
**Working Pressure** *140psi*
**Coupled Wheels** *14 1/2-inch diameter*
**Weight** *1 ton 15cwt*
**Whistle** *LNER streamlined Pacific type*

Having extended his line at Bracknell, Vere Burgoyne was very keen to buy a Bullock 4-6-2, but with the order-book full it would have been some considerable time before his needs were met. However, as Mr Cookson was more than pleased with his slide-valve Pacific No. 2005, with its 14-inch drivers and articulated coupling rods, on his fairly restricted line, he agreed to sell *'The Empress'* back to my father, but immediately placed an order for an identical engine to No. 2005.

*LEFT: It is hard to understand how Bullock produced so many fine locomotives in such a short time, with what must have been fairly basic facilities. No. 3007 is under test, with adjustments being made. The injectors gave some trouble, as can be seen by the amount of water on the ground!*

*BELOW: These were certainly frantically busy days at The Olives. Charlie and Ken are painting one of the new enclosed Pullman coaches on the 'garage' road*

*ABOVE: Alexander Kinloch driving No. 1003 'Western Queen' at Foxhill in 1936. Kinloch preferred 'Western Queen' to the more challenging 'Edward VIII'. Note the plain black livery and tender lettered 'LNER', complete with coal rails* PHOTO: Bob Bullock Collection

*Ken in the driver's seat of 'his' engine, pannier tank No. 3007*

*The tank engines were both delivered by early 1936, but the railway wasn't ready to open to the public until Whitsun*

Within a few days of her return to Farnborough, *'The Empress'* was loaded onto the trailer and taken to Fox Hill. Here she was regularly steamed and used extensively over a period of about ten days, mostly being driven by Father. The locomotive was then re-gauged to 9½-inch in the Prospect Avenue workshops and delivered to Mr Burgoyne in time for the Easter holidays.

It had been hoped that the Foxhill Miniature Railway would also be ready for Easter 1936, but it was not to be. The second tank engine, No. 3008 was now completed and had been delivered to the railway without any difficulties. In addition, the main water tower was now erected at the running sheds. A central-flow vertical boiler was placed next to the water tower, with a steam-driven donkey-pump supplying water to the tank. The boiler also supplied a false draught for raising steam in the locomotives.

This stationary boiler had a rather cranky Ramsbottom-type safety valve, and more than once caused some faint-hearted members of the staff to scatter in panic. It blew off at 75psi, but on occasions the gauge would go beyond 100psi if it were not carefully watched. The cure was to tap the sticky valve lightly to start it, but it seems on one occasion a young painter wiring nearby had been instructed to keep his eye on the gauge and carry out the normal drill, but had momentarily forgotten to check. Whether he over-reacted may never be known, but he discovered the pressure had reached 110psi, and clouted the safety valve with such force that everything went up in the air in a cloud of steam - springs, weights, the lot! The lad fell into the turntable pit in his haste to get away.

There had been numerous runs with all the locomotives up and down the full length of the line to try and arrive at an appropriate timetable, trying various combinations of motive power and rolling stock. The rebuilt 4-2-2, which had been brought over

*ABOVE: This 'works' photo is one of a handful of decent photographs of 'Harvester' as built. Displayed proudly at Fox Hill, the pristine locomotive is still in primer, and has not yet been given nameplates* PHOTO: Bob Bullock Collection

*BELOW: Taking water at the water column near Cove Brook, probably after the opening. 'Harvester' has its nameplate, but no number. She never wore number plates as far as the authors are aware* PHOTO: FHMR Postcard

*An interesting photo of a very early Fox Hill station, showing 'The Empress' on test prior to returning to The Olives for conversion to 9 1/2-inch gauge for Vere Burgoyne's Spring Lanes Light Railway. Behind is 'Western Queen'*

from the workshops just after No. 2006, was quite incapable of hauling more than one loaded coach up the 1:74/1:76 banks, since such a small percentage of its weight was available for adhesion. The little single-driver was double-headed with *'Wendy'* from time to time, but usually worked as station pilot.

'My' engine No. 3007, and the 0-4-2T No. 3008, performed well enough, provided there was sufficient boiler pressure when passing through the loop on the homeward run. The difficulty was that - whereas the Pacifics had an abundance of power to accelerate their trains away after the regulation slack - my small tank engine needed time to 'blow up'. She was usually rostered between two Pacifics - one at Hawley and one at Fox Hill - and this allowed me no time at Lye Copse for this pause for breath. Invariable a Down train would already be waiting to cross and I would have time only to slow down and exchange the token as authority to enter the single-line section.

My father had a lucky escape on one occasion during these preliminary running days. Whilst driving the 0-4-2T with six empty bogies up from Hawley, he approached Fox Hill station miles too fast. It will be remembered that No. 3008 had vacuum brakes, but the stock did not, and when the brakes were applied, they could lock the loco wheels solid. On this occasion, the engine and train slid through the station towards the running shed and turntable, which fortunately was set for his road. He charged over the table, down the length of the engine house and straight through the end of the shed. Finally he emerged smiling, with bits of asbestos sheet all over the front of the engine. He commented to my brother and I what good brakes they were!

One other near thing occurred to No. 3008 whilst I was driving. To gain a road

*Opening day, Whitsun 1936, with Kinloch driving 'Western Queen', a very common scene. This image tells us a lot, but raises some questions. The points are all hand-operated, so the signalbox on the right-hand platform was only working the signals, or simply for show. And the ticket office is at the end of the left-hand platform, which is clearly a bay* PHOTO: Mays of Aldershot Postcard

from the bay platform it was necessary to simply move up to the points, set the points, proceed to clear, reset them, then continue. Having done all these things except the last duty of resetting the points, I made the grave error of leaving my engine in full forward gear instead of bringing the reverser back to mid-gear. Whilst I was walking back to the points the tank started to move forward, there being a slight blow on the regulator. By the time I realised the engine was getting away, it had already got a considerable lead on me. If ever a did a 100 yard sprint it must have been then, and I leapt on the buffers when the loco had gained a speed of about 15mph and brought it under control!

Earlier, *'Black Prince'*, one of the 15-inch gauge Romney, Hythe & Dymchurch Railway's 4-6-2 locomotives, had parted company with its tender near Burmarsh Road whilst the driver had a hand on the regulator. As the engine surged forward, the effect was to open the regulator lever even wider. I heard it was some distance before the locomotive tipped over, preventing what could have been a very serious incident further down the line. Later in the year, Captain Howey asked my father to New Romney for a few days to convert the regulator action to an inward movement to open and vice versa to close.

No. 2573 *'Harvester'* was not quite completed in time for the opening of the railway

*At Hawley the trains initially ran into a single wooden platform, linked by a long walkway past the headshunt to the exit near the Fox Inn. The locomotive then ran round and hauled the train tender-first back to Fox Hill. But here the train is about to depart with the locomotive 'Edward VIII' facing forwards. The shed at the far end of the headshunt is the ticket office*

*The famous closed 'Pullman Cars' at Foxhill solved the problem of getting a quart into a pint pot with lifting (and as in this photo, removable) roof panels, as well as opening doors. This is Fox Hill, but the Pullman has already been lettered SB&CR*

at Whitsun 1936, but there was adequate motive power 'on shed'. All the preparations for running a scheduled service had to be discarded on the first day, as the public began to pile into the coaches as soon as they were emptied. The prototype saloon proved very popular with the older folk.

The signal box was open by this time, and this lent a touch of professionalism to the occasion. Sweets and ice-cream were on sale and the weather was perfect for the carnival-like proceedings. Some passengers bought several tickets at a time, and it was altogether a very successful opening day with everything running perfectly. Takings were about £75 (£4,600 in 2017) from adult fares of one shilling, and child fares of sixpence (£3.06 and £1.53 today).

It was nightfall before the last locomotive had been reversed into the running shed, and the last glowing embers damped down on the track bed. Harvester had developed some cylinder lubrication problems during her initial trials, but this was soon rectified, and she too was entered into service, bringing the available motive power up to seven locomotives.

The layout and position of the stations at Hawley may never be known precisely. Photographs confirm Ken's sketch from the 1987 edition of this book, although the platforms were not islands. This terminus is generally thought to have been eradicated when the loop was built in 1937, but the Railway Magazine article of April 1938 (based apparently on a visit) states: 'Hawley also has a terminal platform and track, run-round road, booking office and signals.' It may be - as suggested in the plan below - that one terminal platform survived

LEFT: H.C.S Bullock about to depart with No. 3008, which seems to have been a successful locomotive, although the vacuum brake needed to be treated with caution!

BOTTOM: Same platform, with 'Edward VIII' departing. It's interesting that the train has a clear signal, but the two inbound signals are also off! Presumably the signalbox (just out of sight on the right) was only staffed when there was more than one loco in steam. The water-crane can't be seen clearly, but it was some distance down the track near the hedge, probably on the right (see page 39), where it was later accessible to trains using the loop too

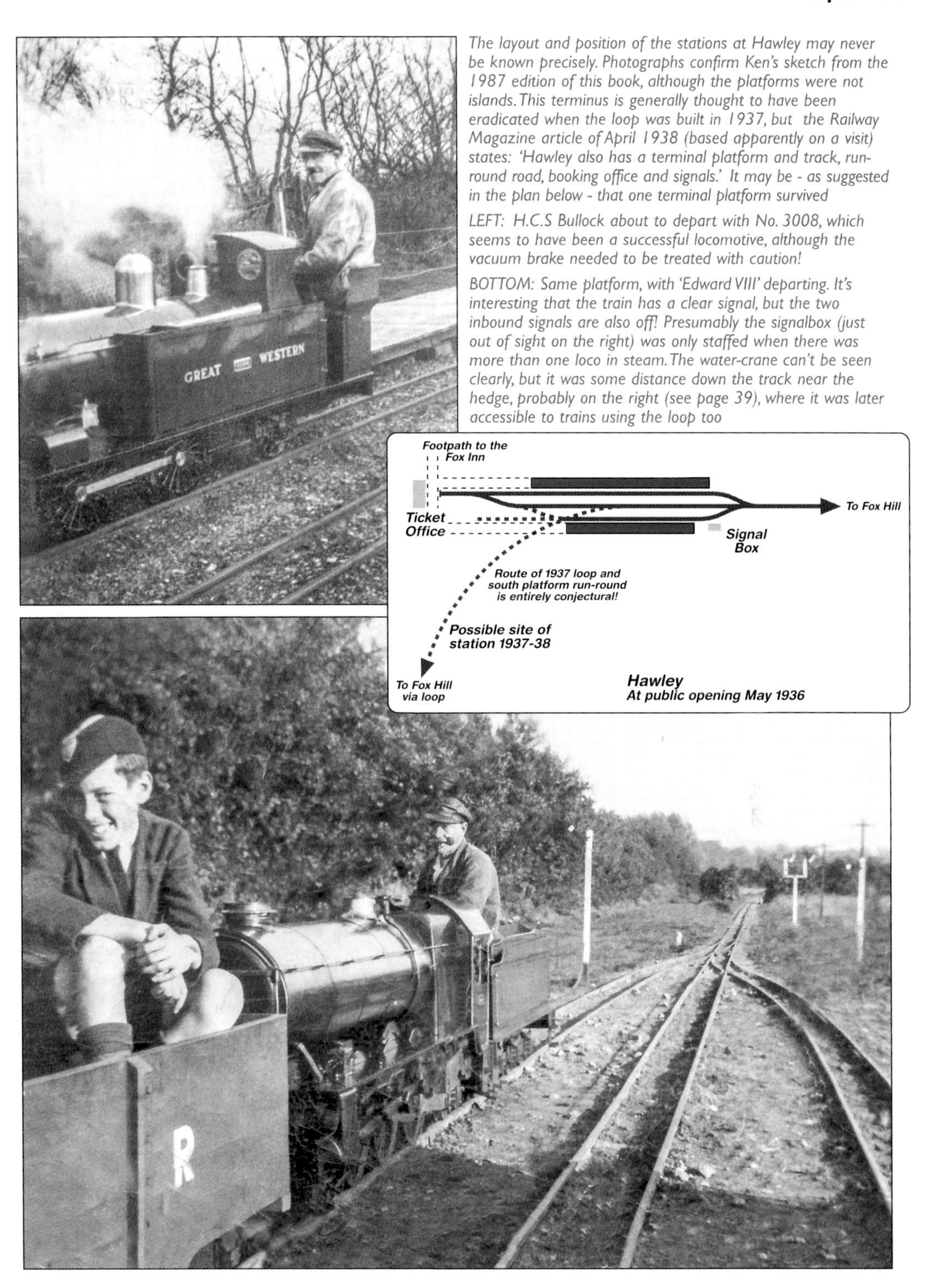

*ABOVE: 'Western Queen' returning tender-first to Fox Hill, probably just east of Lye Copse. Note the 20mph limit sign - few miniature railways today aspire to such a speed, even on the straight! Even fewer local authorities would allow it*

*BELOW: A train of open carriages heading for Hawley, possibly near Cove Brook*

# Chapter 4

*Although not originally Bullock locomotives, both No. 2 'Wendy' (above) and No. 1, the GNR 4-2-2 (below), were rebuilt by Bullock, used at Foxhill, and in 'Wendy's' case, briefly at the Surrey Border & Camberley. 'Wendy', driven by a very cheerful Kinloch, is at Fox Hill as is the GNR single, with Charles Bullock - Ken's older brother - on the right. The single looked the part, but lack of traction meant it was usually confined to shunting duties and publicity*

*Numerous locomotives ran at Foxhill, but it was No. 1003 'Western Queen' that seems to have shouldered much of the work throughout the railway's short life. She was normally driven by Alexander Kinloch, as shown in this storming departure from Fox Hill below, but as above, Bullock drove her too on occasion*

0-6-0T No. 3007 at Fox Hill. Note how the pannier tanks allow both the inside motion and the fittings on top of the boiler to remain accessible

No. 2006 'Edward VIII' ready to depart tender-first from Hawley terminus. Hawley had no turntable, but after the loop was put in, tender-first running ceased to be an issue

No. 1003 'Western Queen', being coaled up at Fox Hill for the run to Hawley in 1937. There is now an all-over station roof, plus the ticket office and cafe in the background. As usual, the train includes a Pullman carriage. This is one of four long ones with three compartments - the shorter version had two. It seems to have drooped slightly! PHOTO: R H G Simpson

# 5
# Problems, Incidents & Accidents 1936-37

*'Coronation' was built in 1936 for Mr Cookson. She is seen above at Fox Hill with a double-bogie tender (three-axle in later photos)*
PHOTO: G Rossiter Collection

While *'Harvester'* was under construction, discussions had been taking place between my father and Mr Kinloch for a large 4-6-2 for the railway. The new locomotive required by Mr Kinloch was to have been after the style of an LMS 'Princess' class, then very much the latest technology. But no final decision had been made on the exact date of commencement, up to the time that Mr Kinloch left for an extended holiday in Switzerland. All of Mr Kinloch's affairs were then left in the hands of his close friend Doctor Bernard.

A week or two after Mr Kinloch's departure, Dr Bernard contacted my father to give him the go-ahead for the new engine. Accordingly, steel for the frames was ordered and cut; axle boxes, horn blocks and brake gear were made, and wheel and cylinder castings were ordered from Wallis & Stevens, and sent to Pantings for machining. Mr Cookson, meanwhile, had been pressing for a completion date for his new Pacific, which resulted in my father working 15 hour days for weeks on end.

When Mr Kinloch returned from Europe to find work had advanced on the new 4-6-2, he immediately put a stop to it, claiming that Dr Bernard had no right to sanction the commencement of the project. The frames, and such of the chassis that had been assembled were removed to the far end of the workshop and covered with dust sheets, this representing a severe financial blow. Had the 4-6-2 been completed by my father it would have been numbered 4010, named *'Princess Margaret Rose'* and finished in maroon livery.

In spite of all my father's efforts now being concentrated on Mr Cookson's locomotive, it was still a bit behind schedule. It was eventually collected by Mr Cookson without ever having been steamed or tested at Prospect Avenue. In fact, the paint was still wet on the buffer planks as we loaded the engine onto Mr Cookson's trailer.

The day after *'Coronation'* left the workshops for Billingshurst, a telegram arrived from Mr Cookson to the effect that all the tests were perfectly satisfactory, with the exception of a troublesome low-pressure injector. The locomotive - although carrying the name Coronation on the splashers - was not fitted with number plates when it left Farnborough. It would have been numbered 4011 (fourth type, eleventh engine). Mr Cookson subsequently numbered it '2011', presumably because it was his second engine currently at Billingshurst.

In spite of the dubious circumstances surrounding the cancellation of the 4-6-2 No. 4010, Mr Kinloch did, however, place a firm order for an Atlantic and work on this started in the summer of 1936. Captain Howey had also been in touch and placed an order for a 15-inch gauge 4-6-2 based on the LMS 'Princess Class'.

The Foxhill Miniature Railway - having evolved from a private test track to a public passenger carrying railway - had now been renamed the 'Farnborough (Hants) Miniature Railway. There was little or no weekday traffic on the railway as autumn approached, so only two engines needed to be steamed daily, usually *'Wendy'* and one of the tank engines.

*H.C.S Bullock could be justifiably proud of this 1936 line-up of locomotives and drivers. Left to right: No. 2006 'Edward VIII', driver H.C.S Bullock; No. 2573 'Harvester', driver 'Dumpy' Edenden (so called because he had lost part of a finger); 0-6-0PT No. 3007, driver Ken Bullock; 0-4-2T No.3008, driver Charles Bullock; 4-4-0 No.2 'Wendy' and GNR 4-2-2 No. 1* PHOTO: Hugh Davies Collection

Maintenance and adjustment of the track was under the supervision of an ex-Southern Railway permanent way ganger. My duties involved collecting the men and their equipment from the section of track they'd been working on. One day I was approached the curve through Lye Copse propelling two trucks with the pannier-tank engine No. 3007 travelling chimney first. As I went onto the curve, I got the distinctly uncomfortable feeling that the superelevation was much too severe.

With the driver's seat on the left side of the bunker, my weight had a further adverse effect on the engine. However, with the works in the hands of such an experienced supervisor, I didn't feel it was the place of a junior driver to criticise the work of a man of such high esteem.

Several of my father's friends and associates had visited the railway at weekends from time to time. Captain Holder had been along, Mr Burgoyne, Mr Cookson, and Captain Howey had also spent a bit of time driving. On the Saturday following the week's maintenance, Mr Richardson, a rather portly gentleman, set off from Fox Hill on the pannier-tank, running light, chimney-first to Hawley, with full tanks and plenty of coal in the bunker. After some 20 minutes, one of the track gang came hurrying up the line to report that the engine had 'gorn over'.

It seems the banking was indeed too severe round the curve, and the weight of the driver had been just enough to tip the engine over - she fell down the bank and landed upside-down in the mud. Fortunately, Mr Richardson had managed to get clear, and was not injured.

My father arrived from The Olives and at first it was thought the engine would have to be dismantled where she lay, but Dad decided to try and get the one-ton Bedford recovery truck from Bradford's Garage reversed down the track, and so winch the loco out. In fact, this was achieved, and within a week or two, No. 3007 was back in service, none the worse for her exploits. Needless to say, the banking was considerably reduced after this incident.

## Bullock 10 1/4-inch Gauge Locomotives

| Date | Number | Name | Wheel Arr | Wheel Size | Working Pressure | Cylinders | Valves | Tractive Effort[1] |
|---|---|---|---|---|---|---|---|---|
| **1932** | **No.1001** | ***The Monarch*** | **4-6-2** | *12"* | *120psi* | *3 1/2" x 5"* | *Piston* | *521lbs* |
| **1933** | **No.1002** | ***The Empress*** | **4-6-2** | *12"* | *140psi* | *3 1/2" x 5"* | *Piston* | *607lbs* |
| **1934** | **No.1003** | ***Western Queen*** | **4-6-2** | *12"* | *120psi* | *3 1/2" x 5"* | *Piston* | *521lbs* |
| **1934** | **No.1004** | ***Mary*** | **4-6-4T** | ***Specification probably as for No. 1003*** | | | | *521lbs* |
| **1935** | **No.2005** | ***Silver Jubilee*** | **4-6-2** | *14"* | *120psi* | *4" x 5"* | *Slide* | *583lbs* |
| **1935** | **No.2006** | ***Edward VIII*** | **4-6-2** | *14"* | *120psi* | *4 1/4" x 5"* | *Slide* | *658lbs* |
| **1936** | **No.3007** | **-** | **0-6-0PT** | *12"* | *125psi* | *2 1/2" x 4"* | *Slide* | *221lbs* |
| **1936** | **No.3008** | **-** | **0-4-2ST** | *12"* | *125psi* | *2 1/2" x 4"* | *Slide* | *221lbs* |
| **1936** | **No.2573** | ***Harvester*** | **4-6-2** | *14 1/2"* | *140psi* | *3 3/8" x 4 3/4"* | *Slide* | *444psi* |
| **1936** | **No.4010** | ***Pr. Margaret Rose*** | **4-6-2** | ***Cancelled, but many parts stored*** | | | | |
| **1936** | **No.4011** | ***Coronation*** | **4-6-2** | *14"* | *120psi* | *4" x 5"* | *Slide* | *583lbs* |
| **1936** | **No.4012** | ***Princess Elizabeth*** | **4-4-2** | *14"* | *120psi* | *4" x 5"* | *Slide* | *583lbs* |
| **1937** | **No.5013** | **-** | **4-4-0** | 9"[2] | *100psi* | *2 1/2" x 4"* | *Slide* | *236lbs* |

Note: Some of these figures are disputed. Those in black are considered reliable, while those in red are speculative

[1] Tractive effort is a theoretical figure assuming 85% boiler pressure [2] Later she was given 12" wheels

# 6
# Triumphs & Disasters
# 1937

*'Edward VIII' departing from Fox Hill station, with Mr Cookson driving*

Evaluation trials and testing of the locomotives were always carried out during quiet periods, usually in the evenings. Harvester was lit up at about 4pm on one such mid-week evening, with the usual assistance from the stationary boiler. With pressure rising nicely in No. 2573, my father took over the duties of attending to the engine, while my brother and I marshalled a very long test train, made up of trucks, open and closed coaches, and the steel ballast wagon.

*'Harvester'* was by now blowing off like mad, and with cylinder drain cocks open, was taken gently forward over the turntable and out beyond the platforms to the signal gantry, where the points were set for her to be backed up onto her extraordinarily long train. After damping down the stationary boiler, and using the last of the steam to lift some water up to the water tower, I climbed aboard the last coach and waved the green light for my father to proceed. After what was undoubtedly a fair degree of slipping, the 4-6-2 got hold of the heavy train and soon began to romp down the descent to Lye Copse, rounding the long curve.

I now had my first clear view of the engine. With just a feather of steam from the safety valves drifting back towards the three saloon coaches in the middle of the train, *'Harvester'* looked really splendid in the setting sun, with her apple green livery in shining contrast to the black smokebox and footplate.

We entered the copse at reduced speed over the first set of points, and I offered the green light as soon as the wheels of the last coach had cleared the points, immediately receiving a 'toot' from the chime whistle, accompanied by the sight of a shower of sparks from the chimney as my father opened his engine out. Accelerating out of the woods, we emerged onto the flat approaches to Cove Brook and beyond, with a burst of speed towards the 1:70 incline towards Hawley, where the facing points controlling the loop were spring-loaded in our favour.

There was barely enough room to run round the train. The locomotive was then replenished with oil, water and coal, while my guard duties were handed over to my brother.

Taking advantage of the favourable road away from Hawley, *'Harvester'* fairly raced along the straight and over the river bridge, with what I thought was a considerable amount of shouldering. Easing his engine down for entry into the darkening woods, my father slowly negotiated the first set of points giving us the right-hand road on the loop. With the engine well clear of the second set of points, he opened the regulator for a good assault on the 1:74 bank. We had only travelled some 50-60yds however, when there was an almighty bang under the footplate of the engine and she began to lurch violently.

My father immediately brought the train to a halt. Examination revealed a bent pony truck axle, possibly caused by a 12lb/yd section of rail on the original track having been fishplated to the newer 9lb/yd rail, leaving a pronounced flange.

We uncoupled half the train and allowed the coaches to run down onto the loop, where we derailed a few of the bogies to prevent a runaway. After the barrier was placed across the line, my father

*4-6-2 No. 2573 'Harvester' raising steam outside the running shed at Fox Hill. A dangerous place for teenage boys to be working unsupervised on locomotives*
PHOTO: Bob Bullock Collection

began to ease *'Harvester'* gently up the gradient to the station approaches, where the shortened train was then uncoupled. The engine was reversed into the running shed, where the fire was drawn and the boiler blown down.

The next day, my brother and I decided to cycle over to the railway to fit the replacement axle ourselves. We detached the injector feed pipes from the tender, uncoupled it and pushed it clear. Then with a car-type screw jack under each side of the engine below the cab, we wound the jacks up together until we had sufficient clearance, removing keeper plates from underneath the axle boxes without difficulty. We then offered up the replacement assembly by lying on the ground and locating the axlebox guides between the horn blocks.

At that moment, the jack on my side slipped and the engine fell back onto the track, trapping me by the arm. My brother's first attempt to lift the rear of the engine with a crowbar only resulted in my arm being partly freed, but I was now trapped by the wrist and in great pain. My brother summoned help on the extension telephone to The Olives, and I can just about remember being dragged clear before passing out.

I came round on my way to Dr Bocket's surgery, where it was discovered that my

*No. 2006 'Edward VIII' passing under the noble signal gantry at the entrance to Fox Hill station. The 1:74 gradient can be clearly seen in relation to the level siding. It was up this bank that the crippled 'Princess Elizabeth' limped home* PHOTO: Bob Bullock Collection

wrist was fractured. In spite of our good intentions, we still had a good ticking off from Dad.

During the course of further trials, *'Edward VIII'* was experimentally fired on small coke. The steaming was not at all impaired, although there was a tendency to lift the fire when working hard. The locomotive proved capable of hauling 12 tons (over 150 passengers, including the weight of the carriages) and was once timed at 40mph on the road. A scale load of 1,820 tons, and a scale speed of 213mph!

The Great Northern single-wheeler No. 1 had proved quite unsuitable for hauling trains. It was therefore decided to mount the locomotive on a flat one-ton platform lorry to advertise the FMR. With the driving axle raised clear of the rails, and the locomotive in steam, the lorry was parked outside such places as the Arcade Cinema in Camberley, the Ritz in Aldershot, and other public places around the district.

Mr Kinloch's new Atlantic, No. 4012 *'Princess Elizabeth'*, was now completed and delivered to the railway late in 1936. The engine was of similar specification to No. 2005, but with cylinders of 5-inch stroke and 3$^{1}/_{2}$-inch bore, slide valves, and finished in blue livery, at the request of Mr Kinloch. The whistle was identical to No. 2006 and it too was placed horizontally, giving the same distinctive tone.

One weekday afternoon, a few weeks after arrival at Fox Hill, *'Princess Elizabeth'* was pushed outside the engine house and made ready for steaming. A train of open coaches were assembled on one of the platform roads, whilst two hundredweight of dry sand was loaded onto a small four-wheel truck. With about 70psi showing on the gauge of the 4-4-2, the three 4-6-2s, Nos. 1003, 2006 and 2573 were hauled out of their respective roads in the engine shed and thoroughly oiled up, the reversing lever of each locomotive being set in the mid-gear position. The three dead engines were coupled to the Atlantic's tender, and hauled forward to the end of the platform. They were then reversed onto the train of coaches (11 in all), the little sand truck, meanwhile, being used to sand about 50 feet of track in front of *'Princess Elizabeth'*.

With my father at the controls, and my brother and me in the sand truck, coupled to the front of the engine, we set off gently from Fox Hill station down the slight inclined stretch beyond the signal gantry. Then came the easier descent down to the curve. In the sand truck we could feel the Atlantic being hunted by the the free-running but heavy 4-6-2s as my father slowed his train for the loop points. After drawing to a halt, the train was reconfigured for the run back. After much pushing and shoving, the sand truck was back in front, with No. 4012 behind running tender first, followed by the three dead engines and the coaches.

The purpose of this particular test, we were informed by my father, was to 'ascertain the haulage capabilities of the 4-4-2 in relation to the locomotive's adhesive weight availability,' which is why we were beginning the test in the dark, gloomy atmosphere of the woods, I suppose, instead of running through to Hawley.

Having applied sand to the wet rails well ahead of the engine, my father drew slowly forward until the last coach was clear of the woods, to allow the barrier to be replaced. Then, with only 30 yards or so of level road before the start of the incline, the Atlantic was opened out in full backward gear and with some degree of controlled slipping, she began to accelerate the heavy train.

From my crouched position in the truck, I was enjoying the sound of the exhaust beat from the hard-working engine, when suddenly it changed to a roar, accompanied by some mechanical noise from the left-hand side of the engine.

After alighting from the tender, my father examined the motion to find a broken expansion link. With the tools always carried on the tender, he removed the radius rod, combination lever and union link from the left-hand motion (see diagram page 113), and with baler twine acquired from the fence, secured the valve in mid-gear.

We uncoupled the coaches from the engine and pushed them gently down to the barrier. Not knowing quite what was going to happen, my brother and I stood by as my father got back on the engine, indicating that we should give the locomotives a push. Our efforts were rewarded with the first exhaust beat from the crippled locomotive as it strained to keep its heavy load moving up the incline on one cylinder.

As the mist rolled down the meadow silhouetting the cattle against the darkening sky, I detected a wry smile as the glow from the fire-hole caught my father's warm features. His engine was holding its own against the 1:74 gradient and the signal gantry was already coming into view. It was a relief to hear the weird exhaust from Princess Elizabeth diminish as she drifted down the platform with her precious load to the safety of the turntable and engine shed.

After the locomotives had been put away, I remember my father driving us home in silence, although no doubt his mind was actively involved in some calculations pertaining to tractive effort.

*Alexander Kinloch's friend Dr Bernard, driving Bullock's' only Atlantic, 4-4-2 No. 4012 'Princess Elizabeth'. The train is running back from Lye Copse over the cattle crossing and about to tackle the stiff climb up to Fox Hill. This was where her test train had to be split after the spectacular failure of an expansion link. It was also where Dr Bernard would later suffer a nasty accident through his own carelessness, contributing to the departure of the Bullock family* PHOTO: FHMR Postcard

# 7
# Last Steamings
# 1937

Mr Kinloch or Dr Bernard would sometimes telephone The Olives requesting an engine be made available on a Sunday, when the line would normally be closed in winter. On one such occasion, Dr Bernard rang for *'Harvester'* to be in steam for his arrival at the line. The locomotive was made ready for mid-morning and driven to the end of the platform, while steam was being raised in the 0-4-2T.

When Dr Bernard arrived, he donned his overalls and set off down the line on *'Harvester'*, running light. When we realised he had left the station, we blew the whistle on No. 3008, but to no avail. We could only wait for the inevitable to occur, as the barrier across the railway line, separating the grazing land from the copse had not been removed!

It was always recognised as the one outstanding rule since the beginning of the railway, that the first driver of the morning removed the barrier, and the last driver at night replaced it.

*H.C.S Bullock's last locomotive was 4-4-0 No.5013, built - rather desperately, one suspects - around the small copper boiler from the vintage ex-Pitmaston Moor single-driver in 1937. Here, H.C.S is running-in the loco on blocks*

Dr Bernard new nothing of this arrangement, as both he and Mr Kinloch usually arrived at and departed from the railway station between these duties being performed.

Sure enough, it wasn't long before Dr Bernard was seen staggering up the track, his face covered with blood. *'Harvester'* had crashed through the bottom of the barrier, breaking the lower of the two stout silver birch poles, but the upper one had dislodged Dr Bernard from the tender. *'Harvester's'* bogie had been derailed in the collision.

As I was the youngest member of staff present, I took the full force of Dr Bernard's wrath. This incident, together with the ordering - then cancelling - of No. 4010, together with other unsavoury aspects at the railway, tipped the scales. The result was that early in 1937, my father removed his two engines, some rolling stock, and sufficient 12lb/yd rail and equipment as he would need to establish a line of his own.

He had come to an arrangement with the proprietor of the leisure complex known as 'California-in-England' to build a line there. The park had been established in 1931, and was less than five miles away, near Crowthorne in Berkshire. Work began immediately clearing the ground, and soon a single-track line was laid half way round the lake, with the station ('Hollywood' naturally) and engine shed adjacent to the boat house.

The line was about 1,100 yards long, and the layout the simplest possible, but this leisure park - so far in advance of its time - was a delightful setting for a railway. The line wound its way round the pine and beech trees, through areas of cultivated rhododendrons offering glimpses of the water lilies and black swans on the lake. After one more easy curve, the track straightened out and crossed an old rustic bridge to reach the terminus. Passengers could alight after a single journey and return on a later train. Some Sundays we were so busy, both engines had to be used double-headed to haul the fully-loaded coaches.

*With the benefit of hindsight, this is a deeply poignant image. The sidings and turntable have gone from The Olives, to be replaced by the ex-FMR signalbox. No. 5013 - the last loco to be built here - is about to set off to the new railway at California-in-England to replace No. 3008, which had been promised to Mr Cookson*

*After the split with Kinloch, H.C.S Bullock chose a good site for his new railway, at the California-in-England leisure park, just five miles to the north, near Crowthorne in Berkshire. ABOVE: Hollywood Station with 0-4-2T No. 3008 about to depart with a well-filled train. It was later replaced by No. 5013, which ran briefly alongside pannier tank No. 3007 until the end of the 1937 season*

*RIGHT: The railway was built incredibly quickly, in just a few months early in 1937. It ran from the well-equipped Hollywood Station to a simple terminus on the other side of Longmoor Lake. The position of this terminus is unclear, but no doubt Bullock's intention was to eventually run all the way round the lake. Hardly any photographs exist, but the map is thought to be broadly correct*

*Hollywood station has a strangely dream-like aura in this image from the 1937 summer season. It was certainly an attractive setting. A rather wistful Ken Bullock is driving No. 3008*

Mr Cookson was keen to buy No. 3008 and a 4-4-0 tender engine was hastily constructed to replace it. The new locomotive was No. 5013 and had outside cylinders with slide valves. It utilised the copper boiler removed from the historic single-wheeler from the Pitmaston Moor Railway.

The 4-4-0 had a six-wheel tender and - perhaps in keeping with the amusement park's American theme - a cow-catcher! The locomotive was painted in GWR colours, and had a copper-capped chimney and brass dome. The latter was partly filled with lead, which added quite a few pounds to assist with adhesion.

In due course, as traffic began to fall off at the end of the 1937 season, No. 3008 was moved to Mr Cookson's line at Billingshurst, and in September it was replaced at California by the 4-4-0. In October, the 0-6-0PT No. 3007 was returned to Prospect Avenue, the 4-4-0 remaining at California under lock and key.

On 22nd October 1937, my father - now physically and mentally exhausted - passed away peacefully at The Olives. As he was laid to rest in the little cemetery behind the

South Eastern & Chatham Railway's Reading to Tonbridge line, a through train from Birkenhead to Ramsgate hauled by a resplendent three-cylinder 'Mogul', drifting down towards North Farnborough Box, urgently whistling up for the outer home signal. It seemed to be a tribute to my father's contribution to the world of miniature railways, a contribution which is never likely to be repeated by one man alone.

*The end of the line, literally and metaphorically. H.C.S Bullock drives No. 3008 onto the little rustic bridge, en route to the eastern terminus of the California-in-England line. After Bullock's death, the railway was re-gauged to 18-inch and ran until the outbreak of the Second War, and reopening afterwards for a few years*

## Publisher's Postscript

*Kenneth Bullock's account ends, naturally enough, with his father's untimely death at the age of just 47. Kenneth, then an impressionable teenager, must have been devastated. He had inherited his father's fascination for mechanical things, and had been very much involved in the locomotive works at The Olives, and the construction of the test-track that became the Fox Hill Miniature Railway.*

*After Bullock and Kinloch parted company, Bullock was completely dependent on the new railway at California-in-England, and he may have found that receipts were less than expected. At the end he owed some back rent (he was paying 30% of ticket sales), and the owner, Mr Cartlidge, seized the track in lieu of payment. He wasn't the only one to make his own arrangements. Bullock died on 20th November 1937, and as soon as news of his death was received, Captain Howey seems to have entered the workshop at The Olives and removed the pannier tank No. 3007, and the frames and other parts from the 15-inch locomotive that had caused some of the cashflow issues.*

*Questions remain as to whether Kinloch paid - or fully paid - for 'Edward VIII' and 'Western Queen'. And there are strong suggestions that he may have quietly removed further items before Bullock's affairs had been properly sorted. In an era when a gentleman's honour still meant something, such behaviour was decidedly ungentlemanly.*

*They were not the only ones. Bullock's erstwhile colleague John Thurston acquired No. 5013 - supposedly secured at California-in-England - together with a complete 4-6-2 chassis, parts, patterns and drawings. Whether he paid the estate, or claimed like others to have been owed money, is not clear, but the manner in which Bullock's property was divided up was - to say the least - unorthodox. And the affair still raises heated discussion 80 years later.*

*He was a brilliant designer and engineer, and - as Ken's touching account shows us - a warm-hearted father too, at least to those of his children who shared his passion for steam. H.C.S Bullock was guilty only of being endowed with a generous and trusting nature.*

*Kinloch went on to run the Farnborough Miniature Railway after Bullock's departure, and with two full-time staff, it's unlikely he ever needed to put in many hours lighting-up and and bedding-down the locomotives. He made many improvements, but there was a certain recklessness about his brief tenure. Operation of the line was improved with the provision of a turning loop at Hawley in the spring of 1937, and some improvements to sidings and covered accommodation at Fox Hill later in the year. He then spectacularly fell out with the local council over the safety and liabilities associated with the footpath crossing near the Cove Brook, a contretemps that Bullock had skillfully avoided.*

*With litigation looming over the crossing issue, Kinloch wrong-footed the council by announcing that the whole line would close and move a mile or two to the Blackwater valley, where the track and stock would be incorporated into his altogether grander (and financially disastrous) Surrey Border & Camberley Railway. In the event, the FMR outlived Bullock by only six months, the final closure coming on Friday 29th April 1938, with the first section of the SB&CR opening the following day. Perhaps Kinloch's haste had something to do with the gathering clouds of war. In the event, the new line was lucky to see two full season's operation before closing for good with the outbreak of hostilities.*

*By this time H.C.S Bullock had been dead for two years, and with the demise of the two railways, followed by five long, grim years of war, and post-war hardship, his contribution to miniature railways was largely forgotten. As was not uncommon at the time, this silence was exacerbated by awkwardness over the manner of his passing, and it would be another 50 years before the genius of H.C.S Bullock would be properly re-examined.*

*Fortunately, his locomotives nearly all survived these fallow years, although survival was in many cases a close-run thing. But thanks to the dedication of those determined to keep the name of H.C.S Bullock alive, we can enjoy most of these machines today.*

*David Henshaw*

*H.C.S Bullock deep in thought beside the Romney Hythe & Dymchurch's 'Hercules' at New Romney, perhaps considering the 15-inch gauge LMS Princess Royal Pacific he was to build for Captain Howey. As we now know, only part of the chassis was completed before Bullock's death in November 1937. Howey quickly removed the completed parts and also the 0-6-0 pannier tank No. 3007 in settlement of the deposit he had already paid*

*After H.C.S Bullock's departure, Alexander Kinloch made several improvements to the Foxhill - now the Farnborough (Hants) - Miniature Railway, but he was already planning bigger things. ABOVE: An intense scene of activity at Fox Hill, with workmen building or rebuilding carriages and other equipment. The passenger carriages have already been renamed 'FCMR', the initials of the then planned 'Frimley & Camberley Miniature Railway'. It became the Surrey Border & Camberley* PHOTO: R G H Simpson

*BELOW: Photographs of the Cove Brook bridge are unusual, and this includes an unusual locomotive too. No. 4012 'Princess Elizabeth' has already been repainted blue, and lettered in readiness for transfer to the Surrey Border & Camberley* PHOTO: Hans Bauman

*The Surrey Border & Camberley opened in April 1938. The line appears many times in the following pages covering the subsequent history of the Bullock locomotives, but as far as the authors are aware, these are the only known colour images of the railway. In May 2016 a collection of Dufay Color transparencies surfaced on an internet auction site. Amongst them were these images of No. 2006 'Edward VIII' at Camberley on the Surrey Border & Camberley Railway in 1938. In the lower photograph, Alexander Kinloch, in peaked cap, can be seen chatting on the platform* PHOTOS: Bob Bullock Collection

# 8
# Locomotive Later History

## 4-6-2 No. 1001 **The Monarch**
### *(later 4-6-4 Audrey, then Bubbles and Whitefire)*

*ABOVE: 4-6-2 No. 1001 'The Monarch' (now renamed 'Audrey') was completed in 1932 for Captain Holder of Keeping, Bucklers Hard, Hampshire. The line is seen here receiving an editorial visit from the 'Model Engineer' in about 1935 (see also page 14). Left to right are Messrs Maskelyne, Captain Holder, his son Terry and Mr Hill, Holder's chauffeur-mechanic. Incidentally, there was a clock over the portico with the legend 'Keeping Time'* PHOTO: Model Engineer

*LEFT: The origins of 'The Monarch's' first boiler are unclear. According to Robin Butterell, it was probably the Goodhand Bros boiler from Captain Holder's dismantled Baltic locomotive. In the 1987 edition of this book it was suggested that Bullock may have built the boiler at Fowler Road, but with the facilities available, this sounds rather unlikely.*

*Whatever its origins, the boiler doesn't seem to have been very satisfactory, and a new copper Goodhand boiler was fitted in 1936, and can be seen being offered up here. 'Audrey' was also given a sand-dome, and the Baker valve gear was replaced by Walschaerts. Much later, the rear pony truck was replaced by a bogie, making the locomotive a 4-6-4. All this work seems to have been carried out at Bucklers Hard*

*During the Second World War, Captain Holder loaned No. 1001 to the War Department to demonstrate to commando troops the most effective use of explosives for destroying enemy locomotives. Fortunately, she survived unscathed!*
*In 1952, the locomotive was bought by Dudley Alexander of Brockenhurst who ran her during the summer months on his circular layout at Meadow End. Here the locomotive - now named 'Bubbles' after one of his favourite ponies - ran an impressive 1,300 miles, earning some £2,000 for charities. In the photo above, Ken Bullock is sitting in the driving seat. Note that the sand-dome is now missing*
PHOTO: Mary Culver

*RIGHT: If this photo is anything to go by, a day out at Meadow End could be pretty entertaining. This looks to be the early 1960s* PHOTO: Bob Bullock Collection

*In 1964 'Bubbles' was acquired by John Fowles, moved to his Stonecot Hill Railway at Carshalton, Surrey and given an overhaul, in which the sand-dome was replaced. The locomotive subsequently moved with Mr Fowles to Hayling Island in Hampshire. Later, 'Bubbles' was acquired by railway author Brian Hollingsworth who was building a spectacular 7¼-inch gauge railway in Wales*

*BELOW: The locomotive was then the subject of a rather unfortunate conversion to 7¼-inch gauge, but at least the Great Western look was restored. Now running once more as a 4-6-2 and repainted in GWR green, 'Bubbles' became 'White Fire'. She is pictured here running at a Heywood Society meeting at Weston Park in Shropshire* PHOTO: Simon Townsend

*ABOVE: In 1994, 'White Fire' was sold to Eastleigh Lakeside Railway, the first Bullock loco to effectively be 'preserved'. Eastleigh was 7¼-inch gauge, but the long-term aim was always to restore her to original condition. She was later given an extensive overhaul, repainted blue, and re-gauged to 10¼-inch, with the railway becoming dual-gauge at the same time* PHOTO: Eastleigh Lakeside Rly

*BELOW: Now restored as No. 1001 'The Monarch', she regularly masquerades as the Reverend Awdry's 'Gordon' for Thomas the Tank Engine events. The driver is Clive Upton* PHOTO: Eastleigh Lakeside Rly

# 4-6-2 No. 1002 **The Empress**

*4-6-2 No. 1002 'The Empress' was built in 1933 for Mr Cookson of Billingshurst, West Sussex. After two seasons, the engine was returned to Farnborough for re-gauging to suit Vere Burgoyne's successive 9 1/2-inch gauge lines at Crowthorne Farm and Spring Lanes near Bracknell (where she is seen above with Mr Maskelyne of 'Model Engineer' magazine driving)*

*RIGHT: Probably taken when the locomotive was first delivered - Vere Burgoyne looks on, while H.C.S Bullock drives. Note the steam-driven feed pump and alterations to the tender top*

*ABOVE: Burgoyne later 'South-Africanised' the engine by painting it yellow, fitting a foreign type of cab and altering the tender top. Later the locomotive was paired with the tender from W.L. Jennings' 'Lake Shore'. The nameplates were removed, with one - rather strangely - being mounted in front of the chimney* PHOTO: John Lucas Collection

*BELOW: After Burgoyne's death, 'The Empress' was bought by Ron Hammett of Bexleyheath, Kent, and run in Danson Park as part of the government-sponsored 'Holidays at Home' scheme during World War 2. At a time when conventional holidays were out of the question, the scheme gave war-weary city families a bit of respite - typically children's entertainments during the day, and dances in the evening. 'The Empress' had been partially de-colonialised by this stage, but she still carries one of her name plates in front of the chimney*

*ABOVE: After several seasons running in Danson Park, 'The Empress' (with her nameplates finally returned to their proper place) was sold to a small company called 'M.E. Locomotives' and run for three seasons in a park in Hornchurch, Essex*

*BELOW: The company later disbanded, and Hammett bought the locomotive back. She was rebuilt by Jack Lambert,and finally restored to her original British appearance*

*LEFT: 'The Empress' was then sold to Barking Council, who ran her in Barking Park until 1962, when - thoroughly worn out - she was replaced with a diesel. Luckily she survived the scrapyard, and was bought by Jim Hutchens who planned to start a steam museum in Ferndown, Dorset*

PHOTO: Bob Bullock Collection

*ABOVE: After the Dorset venture failed, 'The Empress' was stored in a dismantled state for many years, but finally rebuilt by John Goold and Bob Bullock. First steamed at Watford in 1993, she is seen leaving the turntable driven by Ken Bullock* PHOTO: Bob Bullock

*BELOW: 'The Empress', here driven by Rob Hart, is now owned by Eastleigh Lakeside Railway, where she is maintained in immaculate condition and run regularly* PHOTO: Bob Bullock

# 4-6-2 No. 1003 **Western Queen** *(later Tamar Queen)*

*ABOVE: 4-6-2 No. 1003 'Western Queen' was built in 1934 and hired to a Mr Ballance (see also page 21) who used her on his public line at Burnham-on-Sea in Somerset, until Bullock decided he wasn't taking sufficient care of the machine, and took her back to Farnborough for repairs. She went on to play a major role in the construction of the line at Foxhill*

*BELOW: On the opening of the Surrey Border & Camberley Railway, 'Western Queen' was transferred there from the closed Farnborough (née Foxhill) Miniature Railway and paired with Edward VIII's tender*

*ABOVE: Seen here on the turntable at Camberley, 'Western Queen' earned a reputation as a willing and capable locomotive, and she was worked very hard on the SB&CR. She was considered able to pull anything that was put behind her*

*BELOW: Waiting to depart from Farnborough Green station. This photo is interesting in that it shows a nice rake of Foxhill Pullmans with lifting roofs in the centre road, and a rare glimpse of the station facilities, which would put those at many a standard gauge terminus to shame. Sadly, the railway would only survive for two seasons* PHOTO: Bob Bullock Collection

*ABOVE: After closure of the SB&CR, the locomotives went into storage. 'Western Queen' was sold to Charles Lane, who ran her on his short 'Royal Anchor Miniature Railway' in Liphook, Hampshire. Many locomotives passed through Lane's hands, including the famous Wembley Exhibition loco 'Peter Pan'. In c1947, she was sold to Archie Dingle, who ran her at Paignton Zoo, where children put sand down the chimney and ruined the cylinders. Dingle also had a Pacific built by John Thurston which carried many distinctive Bullock hallmarks to the front end and motion, and incorporated some Bullock parts. Confusingly, the Thurston loco was given the nameplates from 'Western Queen', which subsequently became 'Tamar Queen'*

*BELOW: George Woodcock fitted new slide valve cylinders to 'Tamar Queen' after the sand problem at Paignton Zoo, resulting in distinctively raised running plates over the cylinders. She then ran on a temporary line at the South Pier, Lowestoft for two seasons, where she is pictured with a rather unhappy driver!* PHOTO: Bob Bullock Collection

*ABOVE: 'Tamar Queen', was sold to Maurice Densham, who ran her on his impressive garden railway at North Tawton in Devon. The viaduct at North Tawton seems to have been a favourite place for photographs, which is understandable. A spectacular triple header: Guest Atlantic No. 1001 'Sir A. Montgomery', 'Tamar Queen', and Guest Atlantic No. 1002* PHOTOS: Bob Bullock Collection

*BELOW: Another interesting formation, with 'Sir A. Montgomery' and 'Tamar Queen' double heading, and Guest Atlantic No. 1002 just visible as the banker. The long train includes several SB&CR coaches*

*At some point in the late 1950s, 'Tamar Queen' was sold to Lesley Bezley for use on his garden railway near St. Austell in Cornwall. The initials 'OLR' stand for Oakley Light Railway, Oakley being the name of the house. Despite years of hard work and indifferent treatment by a variety of owners, 'Tamar Queen' maintains her dignity* PHOTO: Bob Bullock Collection

*ABOVE: A rare photo of 'Tamar Queen' at work on the OLR*

*BELOW: She was later transferred to Bezley's farm where a new railway was being built, and put into storage awaiting repairs. The boiler was fitted with a new steel barrel, but the original copper firebox was retained. 'Tamar Queen' was eventually reunited with her sister No. 2006 'Princess Elizabeth' (previously 'Edward VIII'), after she too was acquired from Maurice Densham* PHOTO: L Bezley

*ABOVE: 'Tamar Queen' during her long period of storage in Cornwall*

*BELOW: In October 2016, the locomotive was acquired by Eastleigh Lakeside Railway where she awaits rebuilding and returning to steam as 'Western Queen' once more* PHOTOS: Bob Bullock

**Chapter 8**

# 4-6-4T No. 1004 **Mary**

*No. 1004 'Mary' - not the most attractive of the Bullock locomotives - was built for Captain Holder of Keeping in Hampshire, where she ran with the two Atlantics 'Peter Pan' and 'John Terence', and Bullock Pacific No. 1001, now renamed 'Audrey'. She was the last of the 1000-series locomotives, and beneath the side tanks she was mechanically identical to 1001-1003.*

*These photographs don't perhaps show the locomotive in her most favourable light. From the front three-quarters (see page 24) she was a striking and rather elegant large tank engine. In any event, having only one good angle is not enough, and she would go on to suffer a series of indignities and rebuilds* PHOTOS: J A Holder

*ABOVE: The very restricted coal and water capacity caused problems, so a tender was added at Keeping. This addition unfortunately did nothing for her appearance!* PHOTO: J A Holder

*BELOW: 'Mary' left Keeping after Captain Holder's death. The smokebox and chassis were later butchered to create a crude steam-outline petrol-engined 4-4-0 for a line at Happy Mount Park, Morecambe, Lancashire. Luckily, her remains eventually fell into the safe hands of Jeff Price at the Watford Miniature Railway, and are now with an enthusiast who hopes to use the surviving Bullock parts to rebuild her as a steam locomotive in the Bullock tradition* PHOTO: Bob Bullock Collection

# 4-6-2 No. 2573 **Harvester** ***(later Sayajirao)***

*No. 2573 'Harvester' was built in 1936 for the Foxhill Miniature Railway. She was notionally transferred to the Surrey Border & Camberley Railway in 1938, but there is no evidence of her having worked there. In 1941, after closure of the SB&CR, 'Harvester' and three Pullman carriages were sold to the Indian Maharajah of Baroda, Pratapsingh Rao Gaekwad, who built an extensive railway as a present for his son Ranjitsinh's third birthday. The children not only used the railway for entertainment, but travelled to school by miniature railway too!*

*Renamed 'Sayajirao' after the Maharajah's grandfather, the locomotive was destined to work this private railway only briefly. After Indian independence in 1947, the principality was absorbed into Bombay State, and the Maharajah initially became a titular head, but was then deposed in 1951. The railway and its British rolling stock were gifted to the children of the city of Vadodara in 1956, and established on half a mile of the former railway in the Sayaji Baug Gardens, which themselves had been gifted to the city by the Maharajah's grandfather Sayajirao in 1879*

*Some time after 1968, the FHMR Pullman carriages were rebuilt with new doors and without the lifting roof panels, making it impossible for adults to use the train!* PHOTO: Malcolm Shelmerdine

*ABOVE: Over the years, 'Sayajirao' was subtly 'Indianised' - repainted blue and red (yellow too, later) and given some ornate touches, but she clearly remained much loved and well cared for*
PHOTO: Malcolm Shelmerdine

*RIGHT: In 1994, after nearly half a century working in the park, 'Sayajirao' failed a boiler inspection, and was replaced with a diesel locomotive and put on static display in the park. The diesel survived until 2013, when the railway was completely rebuilt and re-gauged, and the vintage rolling stock replaced with the 'Joy Train', a typical Severn-Lamb steam-outline locomotive and narrow gauge rolling stock, which at least allowed adults to use the railway again*
PHOTO: Bob Bullock Collection

*LEFT: In 1998 Ranjitsinh Gaekwad - now the Maharajah - asked for the train to be returned, and 'Sayajirao' was finally handed back in 2003. John Bancroft from the Welshpool & Llanfair Railway was brought in to advise on her restoration, but because of her worn condition, the Maharajah opted instead for a cosmetic repaint in a rather strange shade of green. Described as 'The Flying Scotsman', she now resides in a glass-fronted display case in her original home in the former palace, now a museum. Ranjitsinh Gaekwad campaigned for the Pullman carriages to be returned and displayed with the locomotive until his death in 2012*
PHOTO: Palace Museum

# 4-6-2 No. 2005 **Silver Jubilee**

## *(briefly Princess Marina, later King George VI)*

*4-6-2 No. 2005 'Silver Jubilee' was completed in 1935 for Mr Cookson of Billingshurst, West Sussex, and ran on his private railway. She also operated elsewhere in the county, on miniature railways at Shoreham-by-Sea and - as above and below - in the grounds of Bramber Castle at Steyning. Like many similar lines in the 1930s, this does not seem to have lasted very long*

*ABOVE: From Sussex she returned briefly to Fox Hill, but already lettered for the imminent move to the Surrey Border & Camberley*

*BELOW: This is Farnborough Green on the SB&CR, with 'Silver Jubilee' taking water. She ran here until the line closed in 1939*

*Purchased by Matthew Kerr Senior for his railway at Arbroath in Scotland, 'Silver Jubilee' was renamed 'King George VI' and ran there until 1960, helping the railway to carry a record 55,000 passengers in 1955. In this contemporary postcard, 'King George VI' departs on the right, while steam-outline 'Auld Reekie' is turned* PHOTOS: Matthew Kerr Collection

*Holiday patterns were changing, and in 1960 the railway decided the big Pacific was surplus to requirements. It was sold to W Hammond of Astley Garage, Stourport, reuniting it with several other items of SB&CR stock that had been bought by Hammond from Sir John Lea of nearby Dunley Hall*

*ABOVE: On 8th June 1963, W Hammond sold all of his miniature equipment, the lots including the Bullock Pacifics 'Princess Elizabeth' (formerly 'Edward VIII'), 'King George VI' (formerly 'Silver Jubilee'), and Grimshaw's 'Yankee' 4-4-0 from the Pitmaston Moor Green Railway. 'Princess Elizabeth' was sold to a Mr Phillips of Rock Garage, near Stourport, and 'King George VI' to a Mr Feeney in Melton Mowbray. Sadly Mr Feeney suffered a stroke, and the locomotive ended up languishing in a barn*

*ABOVE: In 1970, "King George VI' was sold in a semi-derelict state to John Fowles who planned to run her at his Stonecot Hill Railway in Surrey, together with sister 'Audrey' (formerly 'The Monarch'), but after the boiler failed an inspection, she was cosmetically restored. After spending a few years on Hayling Island, No. 2005, still in private ownership, went back to mainland Sussex and is seen here on display outside a model railway show in Midhurst in 1980, barely 15 miles from where she first ran*

*ABOVE: 'King George VI' was acquired by Michael Lugg and subsequently returned to Kerr's Miniature Railway in Arbroath, although in this early photo with Matt Kerr 'driving', she is not actually steamable! Mat ordered a boiler, but eventually had to go elsewhere, and the process took some time. Once reboilered, No. 2005 was fully overhauled* PHOTO: Matthew Kerr Collection

*BELOW: As there wasn't really enough work (or indeed track) for such a big locomotive at Kerr's, new operator John Kerr (Matthew Senior's grandson) agreed to her going on long-term loan to Eastleigh Lakeside Railway, where she runs today as 'Silver Jubilee'. Driven here by Matt Archer during the Surrey Border & Camberley weekend in June 2013* PHOTO: Eastleigh Lakeside Railway

# 4-6-2 No. 2006 **Edward VIII** *(later Princess Elizabeth)*

*ABOVE: The enthusiasts who considered the Great Western Railway as towering over its contemporaries had plenty of excuses as to why the GWR failed to build a class of modern 4-6-2s, but one of them actually did something about it, and that was H.C.S Bullock. How well he succeeded is particularly well demonstrated in the illustrations of 4-6-2 No. 2006 'Edward VIII', completed in the winter of 1935-36, and seen here at Hawley, with Bullock driving. No. 2006 followed 'Western Queen' to Fox Hill, where Bullock is known to have regarded No. 2006 as his favourite, and frequently drove this much-admired locomotive*

*BELOW: In 1938, No. 2006 was transferred to the Surrey Border & Camberley Railway, receiving the six-wheeled tender from the ill-fated Atlantic 'Princess Elizabeth'. The engine is seen here on the Camberley turntable* PHOTO: Lens of Sutton

*ABOVE: The turntable road at Camberley seems to have been a favourite spot for photographers* PHOTOS: Simon Townsend Collection

*BELOW: 'Edward VIII' at Farnborough Green. Note the rake of long, slender articulated coaches in the background*

*LEFT: At Camberley station on the SB&CR, now painted black and fitted with a taller chimney*

*After the SB&CR closed, No. 2006 was abandoned, surrounded by tomato plants in the boarded up Farnborough Green station, until 1943, when she was sold to Dunley Hall near Stourport, Worcestershire, and renamed 'Princess Elizabeth'. In 1963, she was auctioned with 'King George VI' at Astley Garage in Stourport, going to a Mr Phillips of Rock, Worcestershire.*

*RIGHT: Maurice Densham later acquired 'Princess Elizabeth', but didn't keep her very long before she was sold to Lesley Bezley to run on his farm railway, where she is seen below, sadly not in steam, with Matthew Kerr at the helm*

*ABOVE: By the time No. 2006 was rescued by the Eastleigh Lakeside Railway in 2007, she was in quite a sorry state, but she was rebuilt by Jesse Moody, and recommissioned as 'Edward VIII' in February 2008* PHOTOS: Eastleigh Lakeside Railway

*BELOW: Bob Bullock driving 'Edward VIII' at Eastleigh's Surrey Border & Camberley weekend in June 2013*

# 4-6-2 No. 2011 **Coronation** ***(later Mighty Atom)***

*4-6-2 No. 2011 'Coronation' was completed late in 1936 for Mr Cookson of Billingshurst, West Sussex. She moved to the Surrey Border & Camberley Railway in 1938, and featured in the official guidebook (above), having been driven by actor Graham Moffatt on the opening day in June 1938. Moffatt was an inspired celebrity choice, best known for his role as a sort of tubby schoolboy in six Gainsborough Pictures comedies playing opposite Will Hay and Moore Marriott. The most famous of these films was the railway-themed 'Oh Mr Porter!', which had been released in January 1938* PHOTO: SB&CR Guide Book

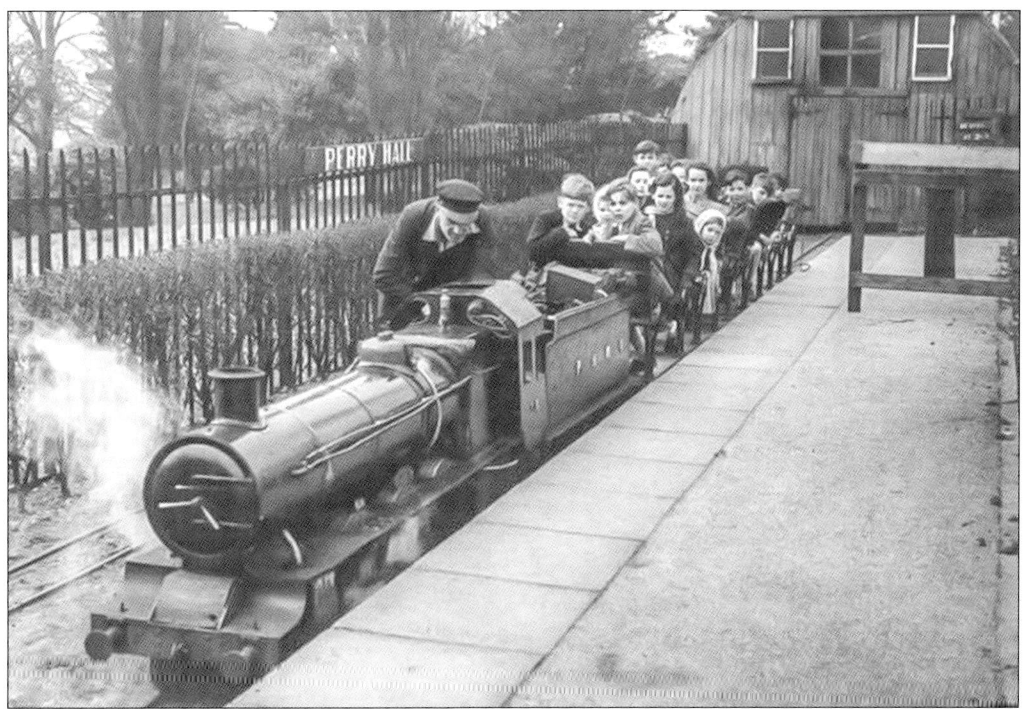


*When the SB&CR closed, 'Coronation' seems to have gone briefly to Vere Burgoyne, then in 1940 to Albert Reeve, who ran her at Perry Hall Park, Perry Barr, West Midlands. The locomotive was renamed 'Mighty Atom', and had her cab modified, as well as being fitted with a taller chimney, which rather spoilt the lines* PHOTOS: (top) Bob Bullock Collection, (bottom) Allan Pratt

*'Coronation' was well cared for at first, but this doesn't seem to have been a very remunerative railway, and by the time it closed in 1955, she was in a pretty dilapidated state. Note the missing buffer and dome cover* PHOTOS: Allan Pratt

*After closure, 'Coronation' was sold to T. F. Cooke, a timber merchant of Wootton Wawen, Warwickshire, then in 1963 to industrialist Dave Murcott, who later retired to Wales, establishing an impressive steam railway in his garden (he had also acquired a Curwen Atlantic). Like many similar lines, the railway was never finished, and the locomotives spent much time in store. In early 2016, 'Coronation' was purchased by Peter Bowers of the Royal Victoria Railway, Southampton, where she now awaits a rebuild to original condition and appearance. She is pictured after arrival in Southampton* PHOTOS: (top) Peter Bowers, (bottom) Bob Bullock Collection

# 0-6-0T **No. 3007** *(later 0-6-0 Firefly)*

*In 1936, Bullock built two small tank locomotives. Both proved to be very useful engines at the Foxhill Miniature Railway, where pannier tank No. 3007 was generally driven by young Ken Bullock, becoming known as 'Ken's engine'. She seems to have been surprisingly capable on some of the significant gradients at Foxhill*

*When Bullock and Kinloch went their separate ways, all the big locomotives remained at Foxhill, so it was No. 3007 and sister No. 3008 that were transferred to Bullock's new railway at California-in-England near Crowthorne early in 1937. In October 1937, as H.C.S Bullock's financial problems grew more serious, she was taken back to The Olives*

*LEFT: In December 1937, just weeks after Bullock's death, No. 3007 was removed by Captain Howey and taken to his Romney, Hythe & Dymchurch Railway in Kent, where she ran on a short 10 1/4-inch line at Dymchurch. This seems to have operated for some time during World War 2 while the RH&DR was closed, and the 10 1/4-inch line was temporarily re-laid on the trackbed of the 15-inch gauge 'main line' in 1945 as Captain Howey struggled to get the railway back on its feet. At some stage, No. 3007 was rebuilt as a tender engine*

PHOTO: Bob Bullock Collection

*After operating briefly at St. Leonards-on-Sea in East Sussex, No. 3007 was sold to Jim Hughes for his new railway at Rock-a-Nore, Hastings, where she was finally given a name - 'Firefly'*

PHOTO: (right), Matthew Kerr Collection

ABOVE: In 1984, 'Firefly' was acquired by Matthew Kerr of Kerr's Miniature Railway in Arbroath. In the late 1980s, Ken Bullock was back driving No. 3007 on a public train, half a century after he last drove at Fox Hill

BELOW: Later, 'Firefly was completely overhauled, and is still running on the line today. She is seen here alongside No. 2005, with Ken in the driving seat of 'his' engine, and Matthew Kerr in charge of the recently restored Bullock Pacific

*'Firefly' driven by Matthew Kerr's son John and piloting 'The Monarch' during the Surrey Border & Camberley weekend at the Eastleigh Lakeside Railway in 2013* PHOTO: Eastleigh Lakeside Railway

# 0-4-2T **No. 3008** ***(later 0-6-0T)***

*0-4-2ST No. 3008 was completed shortly after No. 3007 for the Foxhill Miniature Railway, later accompanying the pannier tank to California-in-England. Mr Cookson bought the locomotive towards the end of the 1937 running season, and operated it for a while on a rather bleak, but quite extensive, miniature railway at Shoreham-by-Sea*

*LEFT: Still at Shoreham, showing the 0-4-2 wheel arrangement. With so much of her weight over the trailing wheels, she must have struggled for grip at times*
PHOTO: Bob Bullock Collection

*BELOW: From Shoreham, No. 3008 was transferred to the Surrey Border & Camberley Railway, where she was rebuilt as an 0-6-0T, which would have greatly improved her capabilities. She is seen here at Farnborough Green driven by Arthur Maxwell. Note the hinged access panel in the bottom of the right-hand water tank*

*BOTTOM: Alexander Kinloch - here departing from Camberley with a lightly-loaded train - is said to have had a soft spot for the little tank engine. After closure of the SB&CR, No. 3008 went to Severn Beach near Bristol, and was rumoured to have been shipped from the nearby Avonmouth Docks to India, but she has disappeared without trace, and nothing is known of the locomotive's fate*
PHOTO: C R L Coles

*No. 3008 at Cove Wood station on the Surrey Border & Camberley*
PHOTO: Peter Mitchell Collection

# 4-4-2 No 4012 **Princess Elizabeth**

*There are few photographs of 'Princess Elizabeth' working at Foxhill (the best are on pages 55, 62 & 108). In a 1937 recreation of Bullock's group photograph, Dr Bernard poses in 'Princess Elizabeth' (left), with Kinloch driving 'Harvester' (right)*
PHOTO: Bob Bullock Collection

4-4-2 No. 4012 'Princess Elizabeth' was Bullock's only Atlantic, built in 1936 to Alexander Kinloch's order. She was basically a shortened version of Bullock's 2000-series Pacifics, paired with a shorter six-wheeled tender. It was apparently intended to use her on the Surrey Border & Camberley Railway and as shown on page 62, she was painted blue and lettered SB&CR while still at Foxhill, but according to Ken Bullock, she suffered a head-on collision with No. 2573 'Harvester', either at Foxhill or the SB&CR. Without Bullock's workshop and expertise, the railway had no facilities for major repairs, but Kinloch had become a director of locomotive-builder Kitson & Co, which solved the problem. Early in 1938, the Atlantic was loaded onto a Bedford lorry together with 'Harvester', the similarly accident-damaged 'Wendy' and the tender tank from No. 2006 'Edward VIII'. The casualties were taken to Kitson's works in Leeds by 'Dumpy' Edenden and Ron Brown, then a junior driver on the new SB&CR.

In due course, both 'Harvester' and 'Wendy' came back, but the boiler and other useable parts from the Atlantic (as well as its number, but not its name), plus the tender tank from No. 2006, were incorporated into an articulated 2-6-0+0-6-2 Beyer-Garratt type locomotive, the first of two built for the SB&CR. The six-wheel tender belonging to 'Princess Elizabeth' was then paired with 'Edward VIII', which would later receive her identity and name plates too (the authors have seen no photograph evidence that the Atlantic ever carried name plates).

The 'new' No. 4012 went into service on the SB&CR, but her performance was not wholly satisfactory. The small carrying wheels at each end were not mounted in pony trucks, but ran in axleboxes in the main frame, giving a long fixed wheelbase. The boiler seems small to supply four cylinders, but she pulled some huge loads at Camberley.

After closure, the Garratt was bought by Charles Lane of Liphook, and later sold to Sir Thomas Salt, who had her 'narrow-gauged' for use on his line in Shillingstone, Dorset. She never worked successfully there, but resided safe and dry in a shed, later being sold and put on display at Wimborne Model Village, before being sold again in 1985 to a Belgian collector. She was eventually repatriated by Peter Bowers of the Royal Victoria Railway, rebuilt (with pony trucks this time), and now runs as 'Basil the Brigadier'

*One of the strangest tales in miniature railway folklore. The Royal Victoria's 'Basil the Brigadier' started life as Bullock Atlantic 'Princess Elizabeth', but in reality very few components remain*
PHOTO: Patrick Henshaw

# 4-4-0 **No. 5013** ***(later Gladstone, then Ivanhoe)***

*No. 5013 was built in 1937 for service on Bullock's new railway at California-in-England, to replace No. 3008, but she was destined to run there for just a few weeks and seems never to have been photographed. Following Bullock's death, she passed through Thurston's hands, and was resold in 1938 to Kerr's Miniature Railway in Arbroath, where she was known as 'Gladstone' but never actually carried nameplates*

*ABOVE: A lovely shed scene at Kerr's from the 1940s. Left to right, Matthew Junior, Matthew Senior, Vic Kinnear, and locos 'Gladstone', 'Silver Jubilee' and steam-outliner 'Auld Reekie'* PHOTOS: Matthew Kerr Collection

*BELOW: Initially, 'Gladstone' proved indispensable, but she was a bit small for hauling the new Pullman coaches, and after the arrival of the bigger engines she saw a lot less use, and was sold in 1947. She is driven here by Matthew Kerr Senior*

*In 1947, No. 5013 followed the well-trodden path to Charles Lane of the Royal Anchor Hotel, Liphook, Hampshire, then in 1953 to George Woodcock of Bishop Stortford, who extensively rebuilt her. She was bought by Peter Hanton from Glossop, Derbyshire in 1954, and ran at both Glossop and Congleton, Cheshire, before being rebuilt with a new copper boiler in 1976 (the original was at least 74 years old by this time!) and entering public service on Peter Hanton's Rudyard Lake Railway in 1987, renamed 'Ivanhoe'. She is being driven (ABOVE) by Ken Bullock*

*RIGHT: No. 5013 was bought by Brian Gent who commenced a major rebuild and had a new steel boiler made by George Mathews. In 2011, the dismantled locomotive was bought by Paul Stileman who completed the rebuild to the very high standard shown here and renamed her 'Gladstone' No. 3009, a number that arguably makes perfect sense*

# Appendix I
# Magazine Features

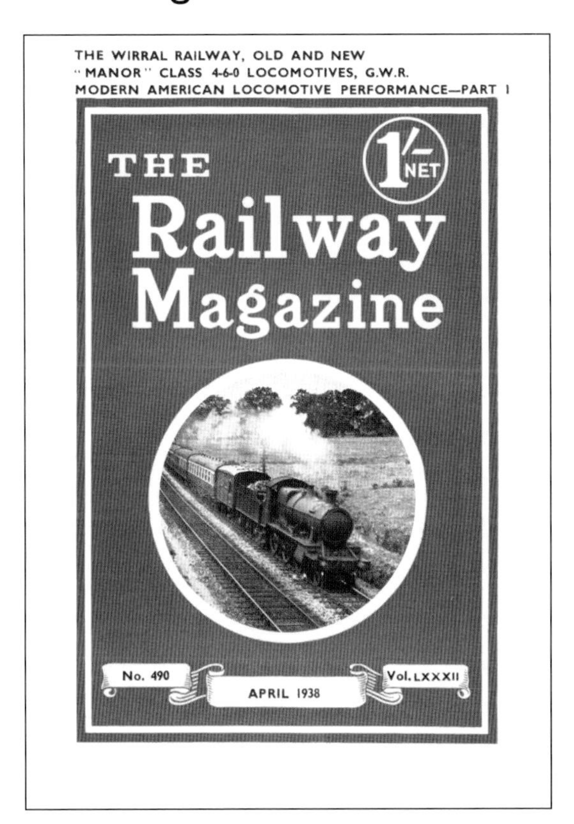
THE WIRRAL RAILWAY, OLD AND NEW
"MANOR" CLASS 4-6-0 LOCOMOTIVES, G.W.R.
MODERN AMERICAN LOCOMOTIVE PERFORMANCE—PART I

THE Railway Magazine

1/- NET

No. 490

APRIL 1938

Vol. LXXXII

## The Railway Magazine
### April 1938
The Farnborough Miniature Railway

Readers of *The Railway Magazine* should find much to interest them in the layout, design, and working of the Farnborough Miniature Railway, a passenger-carrying 10¼-inch gauge line which has been built in some fields close by the road between Frimley and Farnborough.

The main terminus, Fox Hill, consists of three (sic) island platforms, served by five tracks, with two run-round tracks. Behind the platforms are a turntable, running shed and covered carriage sidings (not in picture), while the booking office can be seen above the covered carriage of the train standing in the station.

The terminal layout is correctly signalled, and the signals and points are all controlled from the station signal box. At weekends, when passenger traffic is heavy, the 'signalman' is an attractive young lady. With five engines in steam, the expertness of her control, and her insistence on correct railway procedure from all drivers, are the admiration of the spectators.

Leaving Fox Hill, the line becomes single, but with a long siding for goods and construction wagons. Running along the bottom of a field for about a quarter of a mile, the track turns slightly and there is a cattle crossing. Here the railway enters a 'wild' piece of country, with trees, swamps and bracken.

*The two Bullock tank engines at Fox Hill*

At rather more than half a mile from the terminus, there is a loop, with its own signal box and signals, in communication with Fox Hill box. Leaving the loop (after possibly a short wait for a train coming in the opposite direction) a good speed is attained down a slight gradient and across an impressive bridge over a river. Here is situated a large tank into which river water is pumped for locomotive use. The pipe from this tank feeds the water 50 yards down the line, beyond a set of points, to a convenient place for watering locomotives, whether they are on the Up or Down line. These points are interlocked with, and controlled by, a signal for Up trains.

*This photograph, and the one at the top of page 109 were taken by B. Henri early in 1938 to accompany the Railway Magazine article. This is Bullock Atlantic 'Princess Elizabeth' about to leave Fox Hill - a very rare view* PHOTO: The Railway Magazine

Continuing, the Down line enters upon a large radius circle for turning the trains. At the far end of this circle is situated Hawley station, which also has a terminal platform and track, run-round road, booking office and signals.

The locomotive stud is impressive, being comprised as follows: LNER Pacific

'*Harvester*'; freelance Pacific '*Western Queen*'; freelance Pacific '*King Edward*'; Atlantic type '*Princess Elizabeth*'; 4-4-0 '*Wendy*'; GNR Stirling 4-2-2 (with rather squat chimney), and two Great Western tanks. The railway is owned and operated by Mr Alexander Kinloch, merchant banker of Old Broad Street, London.

*Presumably the young lad was helping Mr Henri with the photography - it certainly makes for a charming image and seems to verify that signals were - or certainly could be - hand-operated, even after the new bigger signalbox had been installed at Fox Hill. Alexander Kinloch is driving 'Harvester'*
PHOTO: The Railway Magazine

*Editor's Note: There is no photographic evidence for more than two island platforms and four platform faces at Fox Hill, although Ken Bullock sketched no fewer than four islands with six platform faces in the first edition of this book, and this Railway Magazine article suggested there were three islands and five faces in those final months before closure in 1938.*

*The layout at Hawley is even less clear. Ken sketched two island platforms in place at the start of public running in early 1936, but the few surviving photographs show two platforms with a centre road for running round. Intriguingly, The Railway Magazine article states quite clearly that Hawley had at least one platform on the loop as well as a terminal platform and run-round road, presumably at some point prior to publication in April 1938. Unless new evidence comes to light, the later layout at Hawley must remain steeped in mystery.*

*'Harvester' in the sidings at Fox Hill. The Railway Magazine article wasn't very comprehensive, but this was a prestigious magazine, already 40 years old, with a circulation at the time of around 25,000*
PHOTO: Mrs Stanley Collection

# The Locomotive Railway Carriage & Wagon Review
## May 1934

Miniature Pacific-type Locomotive
by H.C.S Bullock

The engine illustrated below was built for the $10\frac{1}{4}$-inch gauge by the writer, to his own designs, to produce a machine of maximum haulage power, and at the same time to obtain accessibility to all working parts.

The two cylinders are $3\frac{1}{2}$-inch bore by 5-inch stroke, fitted with piston valves of $1\frac{5}{8}$-inch diameter and $1\frac{1}{4}$-inch travel. The valve timing provides for a steam lead of $\frac{1}{64}$-inch and 75% cut-off in full gear. There is $\frac{1}{64}$-inch exhaust valve clearance or negative lap.

The valves are actuated by a modified Baker gear. The cylinders are lubricated by a mechanical lubricator with separate pump for each cylinder. Drain cocks are operated from the cab. Water pressure relief valves are fitted to the cylinder covers. Inspection covers are provided opposite the four ports, so that the valves can be set by sight instead of the usual measuring on the valve spindle.

The driving and coupled wheels, 12-inches in diameter, are heavy and wide on the treads, whilst the gauge is decreased slightly to enable the engine to run freely around curves of small radius, 35 feet being negotiated with ease.

The bogie wheels are seven-inch diameter, with the inside-frame bogie being

*Bullock doesn't name the locomotive in the article, but this is No. 1002 'The Empress', his second $10\frac{1}{4}$-inch Pacific, prior to delivery to Mr Cookson. The Baker valve gear wouldn't last long, and there would be a few issues with valves, but one senses that Bullock was growing in confidence that he had hit on a winning formula*

allowed 1½-inch lateral play. The rear truck is fitted with outside frames to clear the ashpan, which is sloped forward to allow the ash to drop onto the track; a front damper only is fitted.

The pivot of the rear truck is in front and below the firebox. The wheels are six-inch diameter, and the axle - which at times is heavily loaded - runs on ball-bearings. A powerful steam brake is fitted, and all brake blocks are compensated.

The boiler, which is after the pattern of that of the GWR 4-6-2 *'The Great Bear'*, is 13½-inch diameter at the front end, and at the firebox end, 14½-inch. The barrel is four feet long and the firebox outside is two feet long. Twenty-eight tubes ⅞-inch diameter, and eight flues 1⅛-inch diameter are provided. The grate area is 262 square inches, and the heating surface 37 square feet.

*As Bullock says, the wheels are 'wide on the tread', and slightly narrow to gauge to enable this big machine to negotiate the non-scale curves prevalent on some miniature railways*

The combined safety valve and top-feed is designed in the form of a small steam dome from which steam is collected. This dome is detachable, and provides the inspection cover required by the insurance companies.

The boiler is fed by one injector while standing and by a double-acting pump driven by the crosshead when running. A bypass control valve is fitted and can be operated from the footplate.

The locomotive is ten feet long, 21 inches wide over the footplate, and 30 inches high over the top of the cab.

The tender runs on two four-wheel bogies fitted with ball-bearings, carries 28 gallons of water, and provides ample seating accommodation for the driver. Its length is five foot four inches, making a total length over buffers of 15 foot nine inches. The locomotive is so designed as to be easily converted to the 9½-inch gauge.

*Bullock's early cabs were quite spartan by modern standards. Only one injector, plus a cross-head driven pump, pressure-gauge, reverser and nice accessible regulator*

### *Editor Brian Hollingsworth's notes from the 1987 edition:*

*Bullock says little about haulage capacity in this account, but one reads with awe of a hauling capacity of 12 tons (i.e. a train with 150 passengers), and of 40mph running - a scale 200mph! No doubt the wide treads and deeply-flanged wheels used by Bullock made these astronomical speeds safe, but equally remarkable is the ability of these $10\frac{1}{4}$-inch locomotives to go round curves as sharp as 35-foot radius.*

*George Woodcock, in his book 'Miniature Locomotives', speaks in glowing terms of such Bullock features as the easily get-at-able external regulator, as well as the simple lubricator drive, which substitutes a simple contact arrangement for the usual ratchet-and-pawl. Woodcock also mentions that Bullock did not use superheaters, but this is incorrect as regards the later engines, which had a five-element one, as illustrated. Woodcock rebuilt 'Western Queen', so one assumes No. 1003 and the very similar No. 1002 were not so equipped. But a typical Bullock conundrum is the fact that his account describes a boiler with superheater FLUES, but makes no mention of a superheater!*

Baker valve-gear on No. 1003 'Western Queen'. A lot of untidy linkages, but no tricky-to-machine expansion links

Single-bar crosshead and Walschaerts motion on No. 2005 'Princess Marina' (soon renamed 'Silver Jubilee' and later 'King George VI')

*After experience had been gained, some modifications were found to be necessary. Slide valves were substituted for piston valves, and a second injector for the feed-pump. The first four locomotives (the 1000-series) originally had what Bullock described as 'modified Baker valve gear' instead of the more convention Walschaerts. Baker gear was an American invention which even today has yet to be tried on a full-size locomotive in Great Britain, but in the 1930s its adoption was strongly advocated for little locomotives, by the legendary L.B.S.C in 'Model Engineer' magazine. Its advantage lies in the fact that there are no difficult-to-*

*This interesting photo of one of the later boilers shows the five-element superheater resting in place. Presumably the truck was used to wheel heavy items around at The Olives. The photograph was taken just outside the workshop*

*machine curved slides in the expansion link, only plain pins and bushes. However, in this case there seems to have been a problem, and conventional Walschaerts gear was soon substituted, and would be fitted to all subsequent engines.*

*Another modification, beginning with No. 1004, was the use of single-bar crossheads for the cylinders. In Bullock's day neither this feature, nor the use of outside valve gear, were normal Great Western practise, but he was strangely prophetic, because the last Swindon design - the neat 0-6-0Ts of the 15XX class - had both these features.*

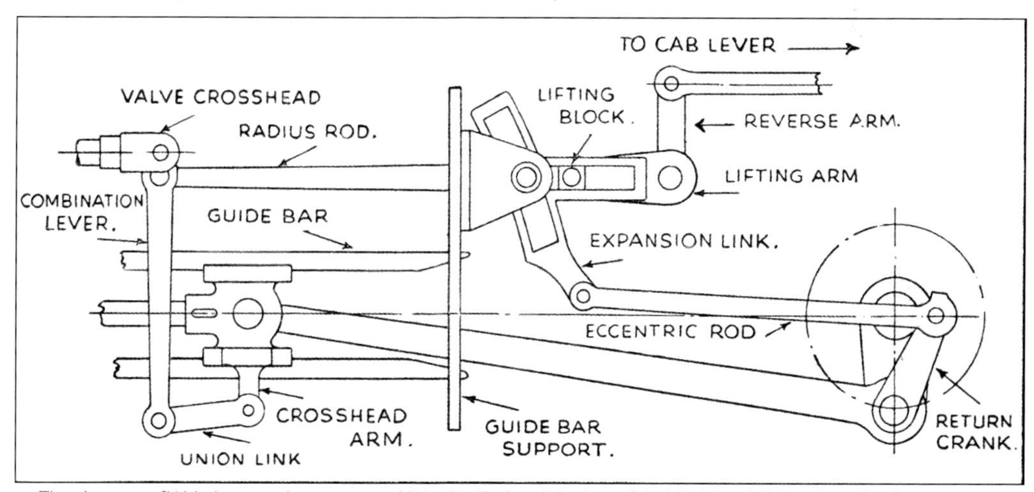


*The elements of Walschaerts valve gear, as published in 'Railway Wonders of the World' in 1935, making the illustration very much contemporaneous with Bullock's work! The difficult machining is in the expansion link, which is carried by the guide bar, and oscillates back and forth, driven by the eccentric rod. The reversing lever lifts or lowers the radius rod via the lifting block so that the expansion link imparts a forward or reverse motion to it, and thus to the valves* PHOTO: Railway Wonders of the World

# Appendix 2
# Non-Bullock Locomotives

*The later history of the Bullock locomotive is complicated by the fact that Bullock designs - and in some cases parts - were incorporated into locomotives built by others. John Thurston had worked for Bullock, and after his death produced no fewer than seven 4-6-2s and 4-6-4s. One of these, confusingly, bore the nameplates 'Western Queen'.*

*The best-known Thurstons were 12¼-inch gauge 4-6-4s with Bullock features - and possibly a few Bullock parts - which long worked the miniature railway at Littlehampton, Sussex. Another maker of Bullock look-alikes was the Motor Gear & Engineering Company, which in 1947 built the 9½-inch gauge 4-6-2 'Princess', and a kit of parts for another similar engine which was built much later by Severn-Lamb and named 'Prince'. This loco now belongs to John Hall-Craggs. These two locomotives were based on 'The Empress' and used piston-valve cylinders. 'Princess' was bought by Ron Hammett and run in Danson Park, Bexley Heath. She was sold to the Bressingham Steam Museum in Norfolk, where she ran until 1994. 'Princess' is now in private hands and undergoing a full rebuild.*

*The Motor Gear & Engineering Pacific 'Princess' running at Bressingham Steam Museum* PHOTO: Bob Bullock Collection

*The 12¼-inch gauge Littlehampton Miniature Railway opened in 1948, operated by two Thurston 4-6-4s, Nos. 1005 and 2010, built in 1947 and '48 respectively. As Ken Bullock (or the editors of the 1987 book) point out rather ruefully, they were built 'with Bullock features - and possibly a few Bullock parts'. They were elegant machines, but very much scale models, without the rugged treads, flanges and bearings used by designers such as Bullock and Greenly. The rear bogies were fitted in place of pony trucks to correct a tendency of the 4-6-2 version to derail in reverse. ABOVE: No. 1005 at Mewsbrook Park station, probably in the 1970s. Alongside is arguably the ugliest steam-outline machine ever built, ex-Bognor Pier No. 3015* PHOTO: Bob Bullock Collection

*Thurston's 'No. 1005' at Littlehampton in 1982, shortly before it was withdrawn after an impressive 44-year working life. Today, both locomotives are owned by J. Hughes of Mablethorpe, Lincolnshire* PHOTO: Patrick Henshaw

*A Thurston 4-6-2 working at Hayling Island in the 1940s. This locomotive is thought to have been called 'Southern Queen', and later made an appearance at Fairbourne, although it never ran there* PHOTO: Bob Bullock Collection

*Believe it or not, this Thurston 4-6-2 - pictured at Worthing in 1947 - would become No. 1005 at Littlehampton. There are detail differences between the various Thurston machines, but the underlying design is the same* PHOTO: Simon Townsend Collection

*In the Bullock context, 'Sir B Montgomery' was the most historically important Thurston machine. It started life as the chassis of the Princess Royal class Pacific cancelled by Kinloch, who said he had never authorised it and refused to pay for it. After Bullock's death it found its way to Thurston's yard, where it was completed, albeit as a slightly ungainly locomotive. It later ran at the short-lived Thames Side Promenade Miniature Railway in Reading, and is now in private hands in Scotland* PHOTO: Peter Scott Collection

*This Thurston 4-6-2 was built in 1946 for Archie Dingle, who ran her in Plymouth for three seasons and swapped her plates with those on 'Western Queen'. Like 'Sir B Montgomery, she also ran briefly at Reading* PHOTO: Peter Scott Collection

# Appendix 3
# Surrey Border & Camberley

*The only known colour photographs of the Surrey Border & Camberley are on page 63. This glimpse of 'Edward VIII' at Farnborough Green is actually a still from an equally rare film. The driver is a Mr. Giles who has treated himself to one of the SB&CR's popular 'Drive a Train' experiences: 'Learn on our big Pacifics with experienced drivers'*
PHOTO: Reproduced from the film 'Ottershaw 1939', Eric Leese Giles [Screen Archive South East at the University of Brighton]

*In this book there are many references to the Surrey Border & Camberley Railway. H.C.S Bullock had little to do with it, but early in 1938 after his death, the Foxhill - later the Farnborough (Hants) - Miniature Railway was superceded by the SB&CR.*

*The SB&CR was a grandiose concept which was cut off on the brink of success by the outbreak of war in 1939. The principal station, Farnborough Green, was located at Frimley Bridges in Surrey (see map page 34), and is quite unrecognisable today beneath the Frimley Business park and motorway slip-roads. The line ran parallel to the old South Eastern & Chatham Railway's Reading-Tonbridge line for a quarter of a mile, then curved away to cross the Blackwater River, terminating, after a run of about two miles, in Camberley near the Blackwater Gas Works.*

*Many of Bullock's locomotives worked on the Surrey Border, and it was a combination of these impressive locomotives, expansive stations, and the proximity to large and entertainment-hungry towns that drew the crowds. Had the line survived it might well have become an attraction on the scale of the Romney, Hythe & Dymchurch, but cruelly cut down by war, it was destined to have a working life of barely two years.*

*The railway was officially opened on 23rd July 1938 by Graham Moffatt (one of the stars*

*LEFT: It wasn't all termini, turntables and flag-waving. The SB&CR had a surprisingly rural character, and although largely double track, there were long stretches of single line through the two woods en route.*

*Here at Watchetts Wood, No. 4011 'Coronation' has just come off the single line from Camberley onto a short double-track section on her way to Farnborough Green. Just out of shot to the right is the Southern Railway's Ascot to Aldershot line, while the Reading to Tonbridge is well within earshot to the left*
PHOTO: Lens of Sutton

*RIGHT: No. 2005 'Silver Jubilee', drifting gently into Farnborough Green. The trackwork on the SB&CR seems to have been simply thrown down in places, and if the line had survived, it would have needed major attention quite soon*
PHOTO: Lens of Sutton

*of that season's big comedy 'Oh Mr Porter!') driving No. 2011 'Coronation'. The line closed at the outbreak of war in September 1939 and - very much in debt - was put into receivership on 10th November 1939, which ultimately resulted in the bank auctioning the equipment, making reopening an impossibility. In war conditions, it's no surprise that the locomotives sold only slowly, but at least they survived being cut up as scrap, a fate that befell many full-size locomotives and several railway lines during the Second World War.*

*No. 2006 'Edward VIII' arriving at Farnborough Green with a well-filled train. The SB&CR opened year round, and although trains were often overwhelmed in fine weather, traffic in the winter could be ludicrously thin. During the summer, trains ran daily until 9.30pm from Farnborough Green, with a 10.05pm return from Camberley. Quite exciting for passengers and drivers alike*

PHOTOS: Surrey Border & Camberley Postcards

*A busy afternoon at Farnborough Green, with No. 2005, driven by Dr Bernard, departing, and No. 2006 'Edward VIII' awaiting departure on the right. Trains ran every half hour at peak times, so locomotive movements would have been almost continuous*

*Both the termini on the Surrey Border were impressive. Camberley had the loco shed, while Farnborough Green had the edge with passenger facilities. Here, No. 2005 'Silver Jubilee' arrives from Camberley at Platform One. In the background is the Reading to Tonbridge line, which must have helped with publicity, although Farnborough North station was actually some distance away. On the other hand, Frimley, on the Aldershot to Ascot line, was just across the road* PHOTO: Bob Bullock Collection

*In 1942, the nation had bigger things to worry about. A train of Foxhill Pullmans waits in vain at Platform 2, Farnborough Green. They were eventually sold to Sir John Lea for use at Dunley Hall, Stourport-on-Severn* PHOTO: Robin Butterell

# Appendix 4
# Modern Times & The Eastleigh Lakeside Railway

*Like all the best stories, this one has a happy ending. In 1987, when Ken Bullock wrote so movingly about his father's life and works, few of his locomotives were in operational condition, some were in a sorry state, and some missing.*

*An impressive line-up of Bullock locomotives at Eastleigh*
PHOTO: Eastleigh Lakeside Railway

*In 1992, Southampton businessman Clive Upton opened the Eastleigh Lakeside Railway to the public. Having had a 7¼-inch gauge railway as a hobby in the grounds of his house, and acquired two locomotives and some rolling stock, it was inevitable that the new railway in Eastleigh Country Park, was built to the same gauge. This was to change when Clive purchased 'White Fire' (originally No. 1001 'The Monarch'), the first of the railway's Bullock locomotives. 'White Fire' had been rather crudely converted to 7¼-inch gauge, and when a major rebuild was required in 2001, she was re-gauged back to 10¼-inch.*

*The railway became multiple gauge 7¼-inch and 10¼-inch - one of just a few in the world -and in 2000, No. 1002 'The Empress' arrived on loan, and was eventually bought by the railway.*

*Although the restoration of 'The Monarch' had upset some enthusiasts, it planted the seed in Clive Upton's mind to establish a collection of Bullock locomotives at Eastleigh, a railway in many ways similar to the ill-fated Surrey Border & Camberley.*

*Eastleigh Lakeside Railway now has a collection of five Bullock Pacifics: Nos. 1001 'The Monarch', 1002 'The Empress', 2005 'Silver Jubilee' (on long-term loan from John Kerr at Arbroath), and 2006 'Edward VIII'. Recently, No. 1003 'Western Queen' has joined the Eastleigh locomotive roster, and will in time be restored. It is true to say that without Clive Upton's enthusiasm and financial backing, the restoration of these beautiful locomotives might never have taken place. They can now be seen in regular action once more on this superb and demanding railway.*

*In June 2013 a Surrey Border & Camberley Railway weekend was organised at Eastleigh. Five Bullock locomotives were running - four Pacifics lettered 'SB&CR' for the occasion - plus No. 3007 'Firefly' visiting from Kerr's Miniature Railway. The locomotives pulled some heavy trains and the event produced sights and sounds not experienced since before the Second World War!*

*With Scottish visitor 'Firefly' at the head, a quintuple-header of Bullock locomotives prepares to leave Eastleigh Parkway during the 2013 Surrey Border & Camberley weekend. H.C.S Bullock would have needed a vivid imagination to conjure up an image like this, on such a railway, some eighty years after he started building 10¼-inch gauge locomotives* PHOTO: Eastleigh Lakeside Railway

*At the Eastleigh Lakeside Railway, Clive Upton (left), with editor Bob Bullock (right), with No. 1002 'The Empress' (left) and Clive's most recent acquisition (right) No. 1003 'Western Queen', currently awaiting restoration* PHOTO: Rob Hart

# Appendix 5
# Kenneth Allan Bullock
## 1923-2006

*Ken Bullock driving the $7\frac{1}{4}$-inch gauge 'White Fire' in 1994, on a new-looking Eastleigh Lakeside Railway* PHOTO: Eastleigh Lakeside Rly

*Ken Bullock was born on 29th December 1923, one of eight children, two boys and six girls. His early years were spent at Fowler Road, Farnborough, where his father was employed at the Royal Aircraft Establishment. Ken soon gained an appreciation of engineering, and of steam locomotives in particular, watching his father build and run many small steam models. After the move to The Olives, Prospect Avenue, Ken was soon helping his father in the workshop, learning to drive and maintain steam locomotives. Eventually, in 1936, he was rewarded with his 'own' locomotive, the Great Western 0-6-0 tank, No. 3007.*

*After leaving school, Ken worked at a local engineering factory, before being called up to join the RAF, serving until the end of World War 2. Several 'dead end' jobs followed, and eventually Ken and a friend decided to emigrate to Australia, to be followed later by Pearl, who was to become Ken's wife. Working on big road construction projects across Australia, and driving heavy trucks, life was hard but enjoyable until Ken and Pearl returned to England in 1957.*

*On his return to England, Ken set about tracing his father's locomotives, following up many leads, most of which proved fruitless, until he discovered No. 1002 'The Empress' working on the Council-owned railway in Barking Park.Ken soon found 'his' tank engine No. 3007 operating at Hastings after being rebuilt as a tender locomotive, and one by one, the other locomotives were tracked down.*

*Unfortunately, none of the Pacifics were operational after 'The Empress' was sold by Barking Council.The others were all in store, mostly in need of major work and with little prospect of running again. 'Harvester' was in India, and No. 3008 had disappeared without trace. However, none of the locomotives had been scrapped, with the exception of 'Mary', of which just some parts remained.The publication by Ken of the first edition of this book, about his father and his locomotives, brought about a resurgence of interest, which culminated in all the remaining locomotives gradually emerging from their resting places, to be rebuilt and eventually put back into operation, or on display.*

*Ken had a wonderful memory for detail and loved to relate stories of the early days at Fox Hill, complete with sound effects... his snifting valves were particularly good! He was able to acquire his father's award-winning model of 'Pendennis Castle' and often drove 'The Monarch', 'The Empress' and 'King George VI' when they were returned to service.*

*Kenneth Allan Bullock passed away peacefully in his sleep at home on 12th May 2006, aged 83. His father's locomotives live on as his lasting legacy, and this is in no small part due to Ken's pride, enthusiasm an dedication.*

*Ken with the rebuilt and regauged No. 1001, restored to the condition in which she left the Fowler Road works in 1930, something that Ken would certainly have remembered* PHOTO: Eastleigh Lakeside Railway

*ABOVE: Ken driving 'The Empress' at the Watford Miniature Railway in 1993, and pausing to study his father's 3 1/2-inch gauge Castle Class 'Pendennis Castle', which he had located and bought* PHOTOS: Bob Bullock

*BELOW: Ken with 'The Empress' again, this time at the Eastleigh Lakeside Railway*

## Publisher's Note:

*Should anyone think Bob Bullock has had an easy task bringing together the surviving photographs, a glance at these before and after images illustrates just how many hours of toil were involved. The photograph shows an unidentified lady driving 'The Empress' at The Olives.*

*A big thank you to Bob for his dedication and hard work, and for dealing good naturedly with my endless nagging for better quality!*

*Numerous other photographs have come to light as we've put this book together, so a follow-up volume looks a distinct possibility. Others are coming to light all the time, and who knows what yet remains to be discovered in dusty Home Counties family albums? H.C.S Bullock has gone, but his legacy will live forever. DAVID HENSHAW*

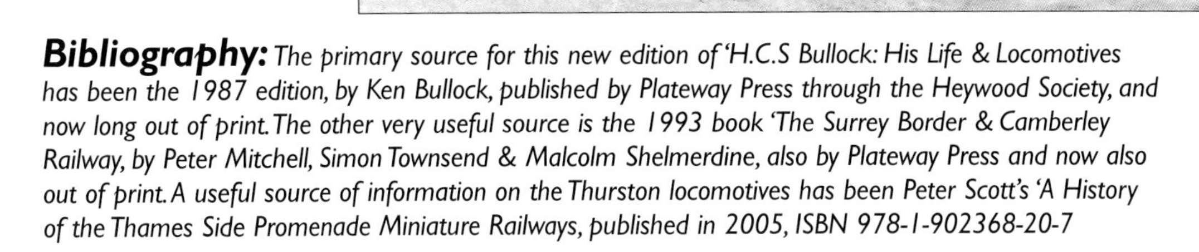

**Bibliography:** *The primary source for this new edition of 'H.C.S Bullock: His Life & Locomotives has been the 1987 edition, by Ken Bullock, published by Plateway Press through the Heywood Society, and now long out of print. The other very useful source is the 1993 book 'The Surrey Border & Camberley Railway, by Peter Mitchell, Simon Townsend & Malcolm Shelmerdine, also by Plateway Press and now also out of print. A useful source of information on the Thurston locomotives has been Peter Scott's 'A History of the Thames Side Promenade Miniature Railways, published in 2005, ISBN 978-1-902368-20-7*

# Index